Introductory Electrochemistry
And Related Applications

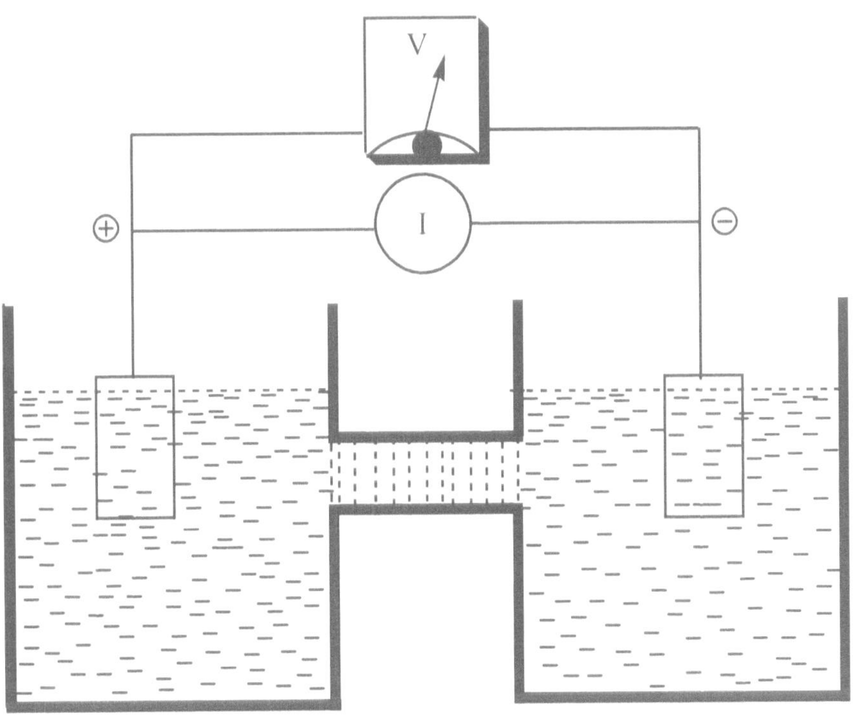

Maurice O. Iwunze

ISBN: 1-4033-9940-9 (e-book)
ISBN: 1-4033-9941-7 (Paperback)
ISBN: 1-4107-8369-3 (Dust Jacket)

This book is printed on acid free paper.

1stBooks – rev. 03/10/08

1	2	3	4	5	6	7	8	9	10	11	12	13	14	15	16	17	18
1 H 1.008																	2 He 4.003
3 Li 6.941	4 Be 9.012											5 B 10.81	6 C 12.01	7 N 14.01	8 O 16.00	9 F 19.00	10 Ne 20.18
11 Na 22.99	12 Mg 24.31											13 Al 26.98	14 Si 28.09	15 P 30.97	16 S 32.07	17 Cl 35.45	18 Ar 39.95
19 K 39.10	20 Ca 40.08	21 Sc 44.96	22 Ti 47.88	23 V 50.94	24 Cr 52.00	25 Mn 54.94	26 Fe 55.85	27 Co 58.93	28 Ni 58.69	29 Cu 63.55	30 Zn 65.39	31 Ga 69.72	32 Ge 72.61	33 As 74.92	34 Se 78.96	35 Br 79.90	36 Kr 83.80
37 Rb 85.47	38 Sr 87.62	39 Y 88.91	40 Zr 91.22	41 Nb 92.91	42 Mo 95.94	43 Tc 98.91	44 Ru 101.1	45 Rh 102.9	46 Pd 106.4	47 Ag 107.9	48 Cd 112.4	49 In 114.8	50 Sn 118.7	51 Sb 121.8	52 Te 127.6	53 I 126.9	54 Xe 131.3
55 Cs 132.9	56 Ba 137.3	71 Lu 175.0	72 Hf 178.5	73 Ta 180.9	74 W 183.8	75 Re 186.2	76 Os 190.2	77 Ir 192.2	78 Pt 195.1	79 Au 197.0	80 Hg 200.6	81 Tl 204.4	82 Pb 207.2	83 Bi 209.0	84 Po 209.0	85 At 210.0	86 Rn 222.0
87 Fr 223.0	88 Ra 226.0	103 Lr 262.1	104 Rf 261.1	105 Db 262.1	106 Sg 263.1	107 Bh 264.1	108 Hs 265.1	109 Mt 268	110 Uun 269	111 Uuu 272	112 Uub 277	113 Uut	114 Uuq 289	115 Uup	116 Uuh 289	117 Uus	118 Uuo 293

57 La 138.9	58 Ce 140.1	59 Pr 140.9	60 Nd 144.2	61 Pm 146.9	62 Sm 150.4	63 Eu 152.0	64 Gd 157.3	65 Tb 158.9	66 Dy 162.5	67 Ho 164.9	68 Er 167.3	69 Tm 168.9	70 Yb 173.0
89 Ac 227.0	90 Th 232.0	91 Pa 231.0	92 U 238.0	93 Np 237.0	94 Pu 244.1	95 Am 243.1	96 Cm 247.1	97 Bk 247.1	98 Cf 251.1	99 Es 252.0	100 Fm 257.1	101 Md 258.1	102 No 259.1

*From ChemGlobe

What do these great men have in common?
Answer: Electricity, Electrochemistry, Energy and Power

André Marie Ampère
(1775-1836)

Charles Coulomb
(1736-1806)

Georg Simon Ohm
(1789-1854)

James Joule
(1818-1889)

Michael Faraday
(1791-1867)

Petrus (Peter) J.W. Debye
(1884-1966)

Erich A.A.J. Hückel
(1896-1980)

Luigi Galvani
(1737-1798)

Walther H. Nernst
(1864-1941)

Count Alessandro Volta
(1745-1827)

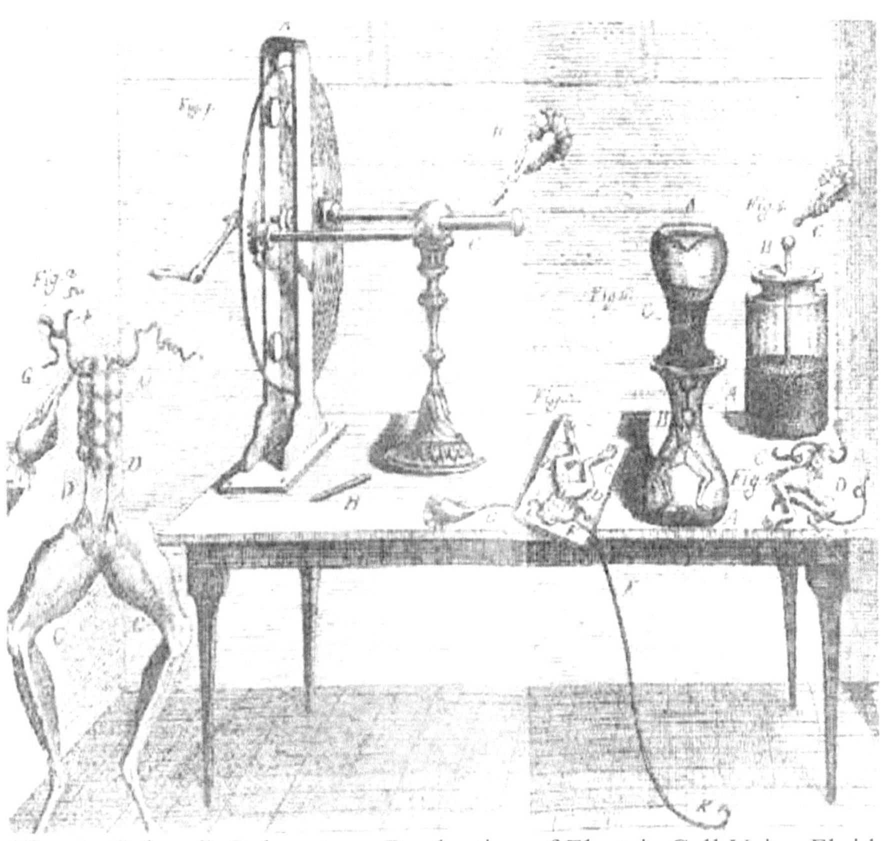

Fig. A. Galvani's Laboratory: Production of Electric Cell Using Fluid
Extracted from the Body of a Frog; Animal Electricity.

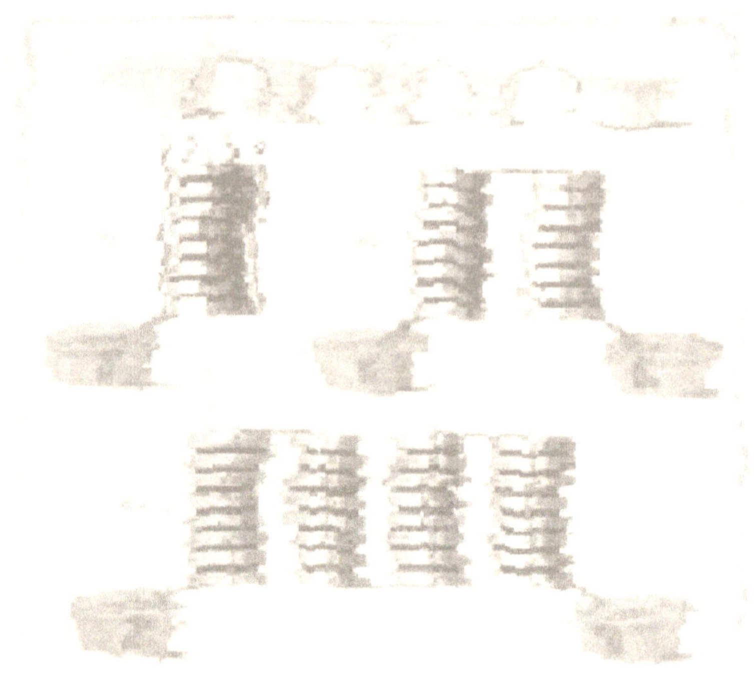

Fig. B. Voltaic Pile of Alternating Metals (Silver and Zinc) to Generate Electricity; The Birth of Battery Technology.

CONTENTS

Prologue

The Curious Beginning of Electrochemistry

During the 1780s, in a series of experiments while at the University of Bologna, Luigi Galvani, an anatomist and physician, observed the contraction of dissected frog muscles, suspended with brass hooks, during a thunderstorm. (Fig. A). This contraction was also observed when there was no thunderstorm but the muscle preparation was in contact with two dissimilar metals. This led him to the conclusion that the frog muscles were generating electricity by themselves, by which he coined the observed electricity "animal electricity". Galvani's experiments established the study of neurophysiology and neurobiology and what was later to be known as bioelectrochemistry.

Alessandro Volta, a good friend of Galvani's, was a physicist at the University of Pavia at the time. He was not convinced of his friend's (Galvani) interpretation of his experiment and the "animal electricity" theory. He believed that electricity could be generated by the contact of two dissimilar metals, and that the frog's nerves were simply electrical conductors. Controversy ensued, and Volta demonstrated his believes by constructing electrical pile by assembling two dissimilar disks of metals, silver and zinc, separated by cardboard disks soaked in brine solution. This pile, known as the "Volta pile" or "Voltaic pile" generated constant current. (Fig. B). This demonstration gave birth to battery technology and its use in the decomposition of water and various other solutions ushered in the science of electrochemistry.

P R E F A C E

This monograph is an introductory text of electrochemistry for students of Science, Technology and Engineering at the very elementary level (Second year University students and their equivalent counterparts in Technical and Teachers Training Colleges). It may also be used as a refresher for upper level students at these institutions. The science of electrochemistry has undergone very dramatic changes in the past fifty years and its slow development at an earlier stage was due mainly to the fact that Nernstian equilibrium treatment of the subject does not tell the whole truth about electrochemical systems and their reactions. Application of kinetics to the study of electrochemistry has accelerated the pace and its eventual recognition as an important branch of science in its own right.

Despite this recognition, elementary textbooks at the first Year University degree program levels have tended to treat this branch of science with cursory and utmost brevity. The unfortunate consequence of this is that students are not fully aware of the important technology that goes along with electrochemistry. These students are, perhaps, aware only of the Nernstian relationship and just the vague notion of accumulator cells.

The aim of this monograph therefore is both to enlighten and to heighten the inquisitive and, I dare say, enthusiastic minds of the young students who may wish to know more about the important fundamentals and application of this branch of science.

Chapters one and two of this short monograph are merely a revision and recapitulation of the basic information. Chapter three discusses the application of the electrochemical theory. Different battery systems, their manufacture and assembly are discussed. Fuel cell technology is also discussed. The very important analytical techniques, potentiometry/ion selective electrode is briefly mentioned. Electrosynthesis of organic compounds, popularized by Manuel Baizer, after his development of electrohydrodimerization of acetonitrile to adiponitrile (a process known as the "Monsanto Process) is discussed. Electroplating and corrosion, important aspects in which electrochemical knowledge is vital, are also discussed.

Application of electrochemical knowledge far transcends these aforementioned fields. They go into electrorefining, electrowinning, electromachining, electrodeposition, electroforming, etc. The discussion of these latter applications are beyond the scope of this monograph and their mention only serves to stress the very importance and extent to which electrochemical knowledge can be applied.

The writing of this monograph was necessitated by the very welcome inquisitiveness of my first year degree students at the Federal University of Technology, Yola, Nigeria, whose ferocious thirst for knowledge was an incentive.

I sincerely hope that their knowledge of this branch of science will be both widened and accelerated for a future pursuit after reading this monograph.

Maurice O. Iwunze
Morgan State University
Baltimore, Maryland

ACKNOWLEDGEMENT

Authors of books cannot claim to be wholly responsible for the production of any given book no matter how small. The author of this monograph is no exception. Therefore, I wish to express my sincere gratitude to the following people who, in no small measure, contributed to the production of the first draft of this monograph: my good friend and colleague, Joe Egila, for his sincere encouragement, Remi Nnodimele who read most of the initial script, Max Orji who typed the manuscript and James Ogbonna, Nahada Allison and Niya Diggs who kindly did all the drawings. My sincere thanks and appreciation go to Janine Nunes who skillfully and meticulously proof read the final script.

My sincere regard goes also to my dear wife for her unending moral encouragement.

MOI

CHAPTER 1

1.0 INTRODUCTION

Electrochemistry is concerned with the study of the mutual inter-conversion of chemical and electrical energies. A system in which electrochemical process takes place has three distinct features as illustrated in Fig. 1. These features are the electrolyte, the electrodes in contact with the electrolyte and the external circuit connecting the electrodes.

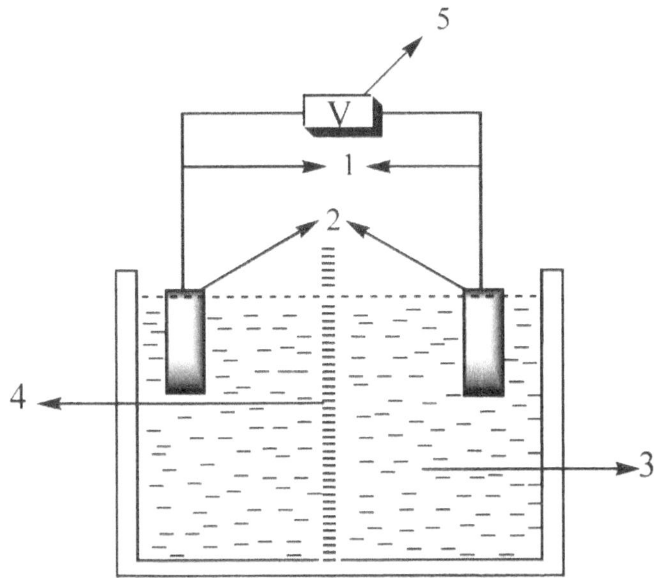

Fig. 1.1 Electrochemical Cell

1 Conducting Wire to Complete the Circuit
2 Metal Conductors (Electrodes)
3 Electrolyte Solution
4 Semi-permeable Membrane
5 Volt Meter

1.1 Electrolytes

Electrolyte is very necessary in any electrochemical process. It is a conductor of the second class. Electrolytes may be defined as those compounds that dissociate in solution to produce ions, that is, negative and positive ions. These ions conduct electricity in solution by their movement. In an electric field, the positive ions migrate to the negative poles and the negative ions migrate to the positive poles. Electrolytes need not necessarily be only those compounds that dissociate in solution. A molten salt that conducts electricity is also an electrolyte.

Electrolytes may be classified by their conductivity in solution. Strong electrolytes are those compounds that dissociate completely in water into ions and are good conductors of electricity. Examples of strong electrolytes and their ionization are as follows:

$$KCl \rightarrow K^+ + Cl^-$$

$$HNO_3 \rightarrow H^+ + NO_3^-$$

Weak electrolytes are compounds that do not dissociate appreciably in solution. They exist in solution mainly in their molecular form. They are poor conductors of electricity. Some known weak electrolytes and their dissociation in water are shown for CH_3COOH, HCN and NH_3.

$$CH_3COOH \rightleftharpoons H^+ + CH_3COO^-$$

$$HCN \rightleftharpoons H^+ + CN^-$$

$$NH_3 + H_2O \rightleftharpoons NH_4^+ + OH^-$$

1.2 Electrolytic Solution

As stated above, when an electrolyte is dissolved in a solvent such electrolyte may dissociate completely, partially or not at all. The degree of dissociation is equal to the ratio between the number of molecules dissociated, n, and the total number of molecules of the electrolyte dissolved in the solvent, N. This may be represented mathematically thus:

$$\alpha = n/N = n/n + n_a.$$

n_a is the number of undissociated molecule. If a solute does not dissociate into ions at all, then $\alpha = 0$, $n_a = N$ and $n = 0$. In this case, the solute may be regarded as a non-electrolyte. An example of a non-electrolyte is sucrose, which exists, in its molecular form in solution. If the ratio of n to N is close to unity, then the solute may be referred to as a strong electrolyte. If $n \ll N$ and $0 < \alpha \ll 1$, the solute may be regarded as a weak electrolyte. The degree of dissociation of an electrolyte is related to the dissociation constant, K. Consider an electrolyte, HX, which dissociates in water, thus:

$$HX \rightleftharpoons H^+ + X^-$$

The dissociation constant may be represented by the well-known equilibrium constant expression:

$$K = \frac{[H^+][X^-]}{[HX]}$$

If we represent the undissociated HX by C_a and the dissociated ions as C^+ and C^- for the cation and the anion, respectively, the resulting dissociation equilibrium constant will be

$$K = \frac{[C^+][C^-]}{[C_a]} \qquad\qquad 1.1$$

The square brackets denote concentration.

However, in the above equation, $[C^+] = [C^-]$. The degree of dissociation, as earlier stated, is α of the total solute concentration, C. Therefore, $[C^+]$ is replaced with αC. And since $[C^+] = [C^-]$, $[C^-]$ will also be αC. It remains now to find C_a in terms of α and C.

$$C = C_a + C^+$$
$$= C_a + \alpha C$$
$$C_a = C - \alpha C \qquad\qquad 1.2$$

Combining equations 1.1 and 1.2 will give equation 1.3

$$K = (\alpha C)(\alpha C)/C(1-\alpha)$$
$$= \alpha^2 C/1-\alpha \qquad\qquad 1.3$$

This equation can be rewritten as

$$\alpha^2/1-\alpha = K/C \qquad\qquad 1.4$$

If $\alpha < 0.1$ or $K/C < 0.01$, then equation 1.4 is simplified to

$$\alpha^2 C = K \text{ and } \alpha = (K/C)^{1/2} \qquad\qquad 1.5$$

Therefore, knowledge of the ionization of an electrolyte (usually a weak electrolyte) can be used to determine its ionization constant as demonstrated in the following example:

Example 1.1 A 0.01 M solution of a certain weak acid, is 0.05 % ionized. What is the K_a of this acid?

Solution Notice that 0.05 % gives an α of 0.0005 (0.05/100). This α value is certainly less than 0.1. Therefore we will use equation 1.5 to quickly solve this problem

$$\alpha = (K_a/C)^{1/2}$$
$$0.0005 = (K_a/0.01)^{1/2}$$
$$(0.0005)^2 = K_a/0.01$$
$$K_a = 2.5 \times 10^{-7} \times 0.01 = 2.50 \times 10^{-9}$$

The degree of dissociation can be obtained through ionic conductance of a given solute. But the dissociation constant, which should be independent of electrolyte concentration at specified temperature and pressure, is obeyed only by dilute solution of weak electrolytes such as acetic acid. What then went wrong?

What went wrong was the fact that the above derivation did not take into account of all the forces and interactions of ions in electrolytic solution. When electrolytes are at infinite dilution their random movement is not impaired, hence the increase in their overall rate of mobility and the consequential high conductivity. If electrolytes are in high concentration, ion-ion interactions occur and ion-pairing also occurs. Such interactions as ion-ion and ion-solvent interactions in electrolytic solution affect the mobility and conductivity of the ions in solution. The effect of these phenomena is a deviation of measured values from theoretically calculated values.

1.3   Activity and Activity Coefficient

In order to make theoretical and experimental values agree, a factor called activity coefficient, γ, is introduced to correct for these shortcomings. For very accurate measurements chemists use activities, a, of ions in solution rather than concentration values. Activity is the effective ionic concentration in solution. The relation between activity and concentration is stated according to the following equation:

$$a_i = \gamma_i C_i \qquad\qquad 1.6$$

where a, γ and C are activity, activity coefficient and molar concentration of species, i, in solution, respectively. Combining this relation with the equilibrium expression given in equation 1.1, one obtains the following more useful relationship, K_{Th}, the thermodynamic equilibrium constant

$$K_{Th} = \frac{[a_{C^+}\gamma_{C^+}][a_{C^-}\gamma_{C^-}]}{[a_{Ca}\gamma_{Ca}]}$$

$$= \frac{[a_{C^+}][a_{C^-}][\gamma_{C^+}\gamma_{C^-}]}{[a_{Ca}]\,[\gamma_{Ca}]}$$

It will also be appreciated that as the concentration of an electrolyte in solution approaches zero (that is, at infinite dilution) the activity of the solution approaches unity. This may be expressed as follows:

$$(a_i \rightarrow C_i)_{Ci=0};\ \gamma_i \rightarrow 1$$

1.3.1   Properties of Activity Coefficient

The activity coefficient is a very important factor in most chemical systems, more so in electrolytic solution that it becomes necessary to characterize some of its properties. These are listed as follows:

a)   activity coefficient depends on the ionic strength of the solution, that is, the total concentration of the ions in a solution
b)   it is a measure of the effectiveness with which the species influences the equilibrium of the ions in solution
c)   ionic strength of the electrolyte in solution determines the value of activity coefficient
d)   for all ions in solution, as ionic strength increases the activity coefficient decreases
e)   activity coefficient depends on the magnitude of the charge on ion not on the kind of ion
f)   neutral molecules are less influenced by the presence of ions in solution, but ions in solution are sensitive to other ions of another electrolyte.

With the foregoing we see the importance of activity coefficient of ions in a solution. These facts notwithstanding, the activity coefficient of individual ions cannot be determined. What is generally determined is the mean activity coefficient of ions of a molecule using the following relationship:

For a compound, X_iY_j, the mean activity coefficient is defined as

$$\gamma_\pm = (\gamma_X{}^i \, \gamma_Y{}^j)^{(1/i+j)} \qquad\qquad 1.7.$$

i and j are charges on the respective ions.

1.3.2   Ionic Strength

We saw above that the activity coefficient depends, in all respects, on the ionic strength of the ions in solution. The ionic strength, μ, of a solution may be determined using the following equation:

$$\mu = \tfrac{1}{2}\sum C_i z_i^2 \qquad\qquad 1.8$$

In this equation, C and z are the concentration and charge of ith ion in a solution. Consistent with the properties of activity coefficient described above, the value of ionic strength depends on the type and charge of an electrolyte. We give below examples of this fact.

Table 1.1 Effect of Charge on Ionic Stregth

Type of Electrolyte	Example	μ
1:1	NaCl	C
1:2	$Ba(NO_3)_2$, Na_2SO_4	3C
1:3	$Al(NO_3)_3$, Na_3PO_4	6C
2:2	$MgSO_4$	4C

Based on the above Table, the calculation of ionic strength of a given solution of different ions can be simplified using equation 1.9.

$$\mu = \Sigma\mu_i \qquad\qquad 1.9$$

Example 1.2 Calculate a) the ionic strength of 0.015 M KCl and b) the ionic strength of a solution made up of 0.01 M KCl and 0.05 M $MgCl_2$.

Solution a) We recall that KCl is a 1:1 electrolyte and using the Table above, its ionic strength is equal to its concentration, that is 0.015. b) KCl is a 1:1 electrolyte and $MgCl_2$ is a 1:2 electrolyte, therefore we use equation 1.9 for this problem.

$$\mu = \mu_{KCl} + \mu_{MgCl2} = 0.01 + 3 \times 0.05 = 0.16$$

The ionic strength of a weak electrolyte may be calculated using the degree of ionization, α, and its concentration, C, as follows:

$$\mu = \alpha C \qquad\qquad 1.10$$

Example 1.3 Calculate the ionic strength of 0.01 M acetic acid, a weak electrolyte. K_a is 1.75×10^{-5}

Solution We will rightly assume that α in this solution is less than 0.1 and therefore use equation 1.5 for its calculation.

$$\alpha = (K/C)^{1/2} = (1.75 \times 10^{-5}/0.01)^{1/2} = 4.18 \times 10^{-4}$$
$$\therefore \mu = 4.18 \times 10^{-4} \times 0.01 = 4.18 \times 10^{-6}$$

1.3.3 Relationship Between Ionic Strength and Activity Coefficient

Activity coefficient of ions in solution may be calculated using the Debye-Hückel relationship given in equation 1.11.

$$-\log \gamma = 0.5085 z^2 \mu^{1/2}/(1 + 0.33\alpha\mu^{1/2}) \qquad 1.11$$

α in this equation is the effective diameter of the ion in angstrom and z is the charge of the ion. For solutions whose ionic strength is less than about 0.01, equation 1.11 can be approximated by equation 1.12

$$-\log \gamma_{\pm} = 0.5085 z^2 \sqrt{\mu} \qquad 1.12$$

Although activity coefficients of most electrolytes are well documented in several textbooks and handbooks, typical values for different ionic species of given effective ionic radius, consistent with equation 1.11 are given below.

Table 1.2 Values of Selected Ionic Radii and their Activity Coefficient at Different Concentrations

Ionic Radii, Å	γ_i at Ionic Strength of *				
	0.001	0.005	0.01	0.05	0.1
Univalent Ions					
9	0.967	0.933	0.914	0.86	0.83
8	0.966	0.931	0.912	0.85	0.82
7	0.965	0.930	0.909	0.845	0.81
6	0.965	0.929	0.907	0.835	0.80
5	0.964	0.928	0.904	0.83	0.79
4	0.964	0.928	0.902	0.82	0.775
3.5	0.964	0.926	0.900	0.81	0.76
3	0.964	0.925	0.899	0.805	0.755
2.5	0.964	0.924	0.898	0.80	0.75
Divalent Ions					
8	0.872	0.755	0.69	0.52	0.45
7	0.872	0.755	0.685	0.50	0.425
6	0.870	0.749	0.675	0.485	0.405
5	0.868	0.744	0.67	0.465	0.38
4.5	0.868	0.741	0.663	0.45	0.36
4	0.867	0.740	0.660	0.445	0.355
Trivalent Ions					
6	0.731	0.52	0.415	0.195	0.13
5	0.728	0.51	0.405	0.18	0.115
4	0.725	0.505	0.395	0.16	0.095
Tetravalent Ions					
11	0.588	0.35	0.255	0.10	0.065
5	0.57	0.31	0.20	0.048	0.021
Pentavalent Ions					
9	0.43	0.18	0.105	0.020	0.009

*Values taken from Lange's Handbook of Chemistry, 4th Ed.

1.4 Electric Conductance

It was earlier stated that one of the properties of electrolytes is their ability to conduct electricity. We use the schematic (Fig. 1.2) to visualize the conduction of electricity in an electrolytic solution.

Consider a strong electrolyte dissolved in a solvent. Being a strong electrolyte, it is therefore assumed to dissociate completely into ions, positive and negative ions.

Under the influence of applied electric field, these ions will migrate to the electrodes of opposite charges as shown.

Since these electrodes are oppositely charged there must be a potential

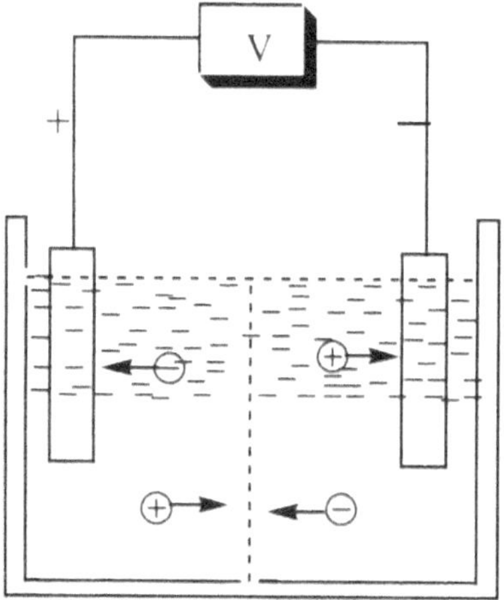

Fig. 1.2 Schematic of the Passage of Electric Current
Through Electrolyte Solution

gradient in the solution. This potential gradient which is the potential difference between the conducting electrodes is simply termed E. The electrolytic solution invariably has a measurable amount of resistance, R, and since there is current, i, flowing through the solution, the entire system must therefore obey the Ohm's Law:

$$i = E/R \qquad\qquad 1.13$$

In electrochemical reactions the desirable electrolyte is the one that is of very high conductance, since that reduces the solution resistance. The conductance of a solution is inversely related to its resistance, thus:

$$L = 1/R \ \text{ or } R = 1/L \qquad\qquad 1.14$$

where L = conductance in mho (Siemens) and R = resistance in ohms. It is easy to see the connection between solution conductance to a desirable electrochemical parameter by considering the Ohms Law shown in equation 1.13

Combining equations 1.13 and 1.14 one obtains equation 1.15, thus:

$$E/i = 1/L \text{ and } E = i/L \qquad\qquad 1.15.$$

Equation 1.15 shows that electrochemical potential is inversely proportional to the conductance of the solution. The implication of this is that at high solution conductance the resultant electrochemical potential is reduced, a factor that is useful in electrolytic reactions.

It is conventional in solution chemistry to report molar conductivity, Λ, for electrolytes. Molar conductivity is a function of the measured molar specific conductance, κ, and the concentration, C, of the electrolyte, thus:

$$\Lambda = \kappa/C. \qquad\qquad 1.16$$

κ has the unit of $(\Omega\text{-m})^{-1}$ and C is expressed in mol/m^3. Therefore, the unit of Λ is m^2/mol or in the current SI system, Λ is specified as Siemens-m^2/mol or simply as S-m^2/mole. Λ values for weak electrolytes approach those of strong electrolytes at sufficiently dilute solution, that is, at infinite dilution.

Extensive published work done by Friedrick Kohlrausch and his coworkers between 1869 and 1880 on conductivity experiments gave useful information regarding molar conductance and their chemical applications. Among these are:

a)   A plot of Λ Vs $C^{1/2}$ is linear giving an intercept of Λ_o, known as conductivity at infinite dilution, that is:

$$\Lambda = \Lambda_o - kC^{1/2} \qquad\qquad 1.17$$

b)   The degree of ionization, α, can be obtained using the relation

$$\alpha = \Lambda/\Lambda_o \qquad\qquad 1.18$$

This is analogous to the relation derived for weak electrolytes in Section 1.2

c)   For a given electrolyte Λ_o is a contribution of the cation and the anion of the electrolyte, that is

$$\Lambda_o = \Lambda_o^{+} + \Lambda_o^{-} \qquad\qquad 1.19$$
$$= \lambda_{o,i}^{+}C_i + \lambda_{o,i}^{-}C_i$$

This is also known as the Kohlrausch Law of Independent Migration of Ions.

The consequence of the last application is that it is possible to calculate the Λ_o of weak electrolytes if the values of their salts, which are strong electrolytes, are known as can be seen from the following example:

$$\Lambda_o(HAc) = \Lambda_o(NaAc) + \Lambda_o(HCl) - \Lambda_o(NaCl)$$

However, a careful examination of this relation reveals that these species are strong acids and strong bases and salts derived from them. We can therefore derive the used species as follows:

$$NaOH + HCl \rightarrow NaCl + H_2O \qquad (i)$$
$$NaOH + HAc \rightarrow NaAc + H_2O \qquad (ii)$$

Since in the equation no water was given, we therefore reverse equation (ii) and sum the result, that is,

$$NaOH + HCl \rightarrow NaCl + H_2O \qquad (Neutralization\ Reaction)$$

$$NaAc + \underline{H_2O \rightarrow NaOH + HAc} \quad \text{(Hydrolysis Reaction)}$$

Sum: $\quad HCl + NaAc \rightarrow NaCl + HAc \quad \text{(iii)}$

We assume that because of the nature of these reactions, the following relation holds true:

$$\Lambda_{o\ products} = \Lambda_{o\ reactants}$$

Therefore, $\Lambda_o(NaCl) + \Lambda_o(HAc) = \Lambda_o(HCl) + \Lambda_o(NaAc)$

And $\Lambda_o(HAc) = \Lambda_o(HCl + \Lambda_o(NaAc) - \Lambda_o(NaCl)$

Notice that this last equation is the same as equation (iii) above. We list in Table 1.3 molar conductivities of selected ions. An xtensive list can be found in most reference and handbooks.

Table 1.3 Equivalent Ionic Conductances (Ω^{-1} cm^2 equivalent^{-1})*

Cations	λ_0	Anion	λ_0
H^+	349.99	OH^-	198.4
Li^+	38.7	F^-	55.4
Na^+	50.12	Cl^-	76.39
K^+	73.54	Br^-	78.18
$Rb+$	77.85	I^-	76.88
$Cs+$	77.3	NO_3^-	71.50
$1/2Ba^{2+}$	63.6	HCO_3^-	44.5
$1/2Ca^{2+}$	59.53	$HCOO^-$	54.6
$1/2Mg^{2+}$	53.0	$C_6H_5COO^-$	32.4
$1/2Pb^{2+}$	69.5	$1/2SO_4^{2-}$	80.0
$1/2Cu^{2+}$	53.6	$1/3Fe(CN)_6^{3-}$	100.9
$1/3La^{3+}$	69.7	$1/4Fe(CN)_6^{4-}$	110.6
Ag^+	61.9	CH_3COO^-	40.9
Tl^+	74.7	CH_2FCOO^-	44.4
NH_4^+	73.5	CH_2ClCOO^-	42.2
$CH_3NH_3^+$	58.7	CH_2BrCOO^-	39.2
$(CH_3)_2NH_2^+$	51.9	CH_2ICOO^-	40.6
$(CH_3)_3NH^+$	47.2	CH_2CNCOO^-	43.4
$(CH_3)_4N^+$	44.9	$CH_3CH_2COO^-$	35.8
$(C_2H_5)_4N^+$	32.6	ClO_3^-	64.6
$(C_3H_7)_4N^+$	23.4	BrO_3^-	55.7
$(C_4H_9)_4N^+$	19.4	IO_3^-	40.5
$(C_5H_{11})_4N^+$	17.4	ClO_4^-	67.3
		IO_4^-	54.5

* Data were taken from *Electroanalytical Chemistry Basic Principles and Applications* by James A. Plambeck, Wiley, New York, 1982

1.5 Transport Number and Diffusion Coefficient of Ions

Transport number and diffusion coefficient are importantly useful in electrochemical studies. Both of these parameters are obtainable from a knowledge of molar conductivities.

Transport number, t_i, which is also known as transference number, describes the fraction of current carried by ith ion in solution, that is

$$t_i = \lambda^o_i/\Lambda_o \text{ and } t_i^{+} + t_i^{-} = 1 \qquad\qquad 1.20$$

Arising from ionic mobility, u, is the Nernst relation between this quantity and the diffusion coefficient, D, of ions in solution. This relation shown in equation 1.21 is also seen to be derived from molar conductivity

$$D_i = RT\Lambda_i/ z_i^2 F^2 \qquad\qquad 1.21$$

z_i is the charge of the ith ion and F is Faraday's constant.

1.6 Electrodes

Electrodes are metal conductors which are in contact with electrolytes and their ions in solution. They serve as means of exchange of electrons with the reactants in solution and also as a means of transfer of electrons to the external circuit. In most electrochemical systems, two electrodes are necessary. These are the anode and the cathode.

The anode is the electrode at which oxidation takes place. The part of the electrolyte in contact with the anode compartment is usually referred to as the anolyte.

The cathode is the electrode at which reduction takes place. The electrolyte at the cathode compartment is referred to as the catholyte.

The external wire, referred to as the conductor of the first kind, connects the anode and the cathode to a measuring device, the voltage meter. Electron exchange between these electrodes is carried through this conductor.

As the electricity is conducted by the ions in solution, the circuit is completed, by electron transfer, through the wire of the external circuit. This concept will be made clearer in the next chapter using Fig. 2.1. The electrodes mentioned above are metal indicator electrodes and they are classified in accordance with the mechanism of their response to electroactive materials in the solution.

1.7 Kinds of Metal Indicator Electrodes

There are four main kind of metal indicator electrodes: electrode of the first kind, electrode of the second kind, electrode of the third kind and the redox electrode.

1.7.1 Electrode of the First Kind

Electrode of the first kind could be metal or amalgam electrodes in contact with its ions in solution. In this kind of electrode, the potential is controlled by the activity or the concentration of the ionic species in solution. An example of this kind of electrode is the silver metal in a solution of silver nitrate.

The reaction is:

$$Ag \; \rightleftarrows \; Ag^+ + e^-$$
$$E_{ind} \; = \; E^o_{Ag+/Ag} - 0.059 \log a_{Ag+}$$

1.7.2 Electrode of the Second Kind

This kind of electrode consists of a metal-insoluble salt or oxide or hydroxide immersed in a solution containing the anion of the sparingly soluble compound of the electrode. An example of the electrode of the second kind is the silver-silver chloride electrode

$$
\begin{array}{ll}
AgCl & \rightarrow \; Ag^+ + Cl^- \\
\underline{Ag^+ + e^-} & \underline{\rightarrow \; Ag} \\
AgCl + e- & \rightarrow \; Ag + Cl^-
\end{array}
$$

The potential for this cell is:

$$E_{ind} = E°_{Ag+/Ag} - 0.059 \log a_{Cl^-}.$$

It can be seen that the potential of the electrode of the second kind, such as shown above, is controlled by the anion of the sparingly soluble electrode.

1.7.3 Electrode of the Third Kind

The electrode of the third kind involves a redox couple of the electrode of the second kind and two solutions equilibria. Consider an electrode of the second kind, say $PbCO_3(s)$. The reaction with respect to this electrode may be represented as:

$$PbCO_3(s) \rightleftarrows Pb^{2+} + CO_3^{2-}$$
$$\underline{Pb^{2+} + 2e^- \rightarrow Pb}$$

$$PbCO_3 + 2e^- \rightarrow Pb + CO_3^{2-}$$

The cell potential may be written as:

$$E_{ind} = E°_{Pb2+/Pb} - 0.0295 \log a_{CO3}^{2-} \qquad (a)$$

If at this point, a small solution of Mg^{2+} is introduced into this system, we have another solution equilibrium, thus:

$$Mg^{2+} + CO_3^{2-} \rightleftarrows MgCO_3(s)$$

$$K_f = \frac{1}{a_{Mg2+} a_{CO3}^{2-}}$$

We can write the formation constant, K_f, of $MgCO_3$ in terms of the activity of the carbonate ion thus:

$$a_{CO3}^{2-} = \frac{1}{a_{Mg2+} K_f}$$

Inserting this relation into the equation (a) above we obtain:

$$E_{ind} = K - 0.0295 \log 1/K_f - 0.0295 \log 1/a_{Mg2+}$$
$$= K + 0.0295 \log K_f + 0.0295 \log a_{Mg2+}$$

It can be seen from this final relation that the electrode potential of this electrode is dependent on the activity of the Mg^{2+} ion, thus making the Pb electrode an electrode of the third kind for calcium.

1.7.4   Redox Electrode

This kind of electrode consists of an inert metal in contact with ion that can exist in two different oxidation states. A typical example of this kind of electrode is the Fe^{3+}/Fe^{2+} couple in a solution with an inert platinum electrode, that is $Pt/Fe^{2+},Fe^{3+}$. The reaction is as follows:

$$Fe^{3+} + e^- \rightleftharpoons Fe^{2+}.$$

The electrode potential will then be:

$$E_{ind} = E^o{}_{Fe3+/Fe2+} - 0.059 \log a_{Fe2+}/a_{Fe3+}.$$

1.8 Definition of Some Electrochemical Terms

In order to fully understand and appreciate the discussions that follow hereafter, it will be appropriate at this point to define some important electrochemical terms.

i)   Ampere, Amp or A, is a unit of current whose unit is Coul/s

ii)   Cell Voltage, V, is the overall voltage resulting from an electrochemical reaction.

iii)   Coulomb, Coul., is the quantity of electric charge. Its unit is amp-sec.

iv) E^o is the potential of an electrode reaction measured versus the Standard Hydrogen Electrode (SHE) when the activity or concentration of the reductants and oxidants exchanging electrons in solution is unity. Its unit is Volt.

v)   EMF is the electrode potential measured at Standard conditions, that is at unit activity and 25 $^{\circ}$C.

vi)   Faraday, F,   is the quantity of electricity per equivalent of chemical change. Its value is approximately 96500 Coul

vii)   Half cell is the compartment where either oxidation or reduction takes place.

viii)   Half reaction is a reaction occurring at an electrode in a half cell.

ix)   Salt bridge is the medium through which ionic contact is made between two separated solutions. The bridge could either be in the form of gel such as saturated KCl in an agar gel or solution of NH_4NO_3 or saturated KCl.

CHAPTER 2

2.0 ELECTROCHEMICAL REACTION

An Electrochemical reaction is a reaction in which the reacting species exchange electrons not directly but through an interface. The interface is the phase between the solution and the electrode. Before a detailed discussion on this subject can be made the fundamental knowledge of Redox reaction is in order.

2.1   Redox System

A redox system is a system comprising of two species. One is the reductant and other is the oxidant. In other words, we are talking of an oxidation-reduction reaction. In this type of reaction the oxidizing agent gains some quantity of electrons which is lost by the reducing agent. In an oxidation-reduction reaction the reducing agent is itself oxidized and the oxidizing agent is itself reduced. To illustrate this, consider a redox system:

$$Ce^{4+} + Fe^{2+} \rightarrow Ce^{3+} + Fe^{3+}$$

This reaction states that Ce^{4+} is oxidizing Fe^{2+} to Fe^{3+} and during this process Fe^{2+} is losing an electron to become Fe^{3+}. In this same process the Ce^{4+} which is oxidizing the Fe^{2+} is itself being reduced by accepting one electron to go from Ce^{4+} to Ce^{3+}. The entire process may simply be stated that in any oxidation-reduction processes the number of electrons lost by a species must be equal to the number of electrons gained by the other species in the solution.

2.2 Balancing Oxidation-Reduction Equations by "Half Reaction Method"

This method may also be called the "ion-electron" method. Whatever the name is, the cardinal point in balancing a redox reaction is that the equation is balanced both atomically and electronically, that is, the charge on both sides must balance and atoms of each kind must balance on both sides of the equation. The following is a useful procedure for balancing a redox equation:

a) Write a skeleton equation with the formula of the reactants and products
b) Examine the skeleton equation and determine which elements change oxidation states
c) Using coefficients, balance the electron gained or lost in the equation
d) Balance all other elements in the equation
e) Balance each reactant with its product separately
f) Multiply each of the half reactions with appropriate coefficient(s) to balance the electron or charges on both sides of the equations.
g) For reactions in acidic solution H^+ is written on the left with H_2O written on the right hand side of the equations
h) In basic solution OH^- is written on the right with H_2O written on the left side of the pertinent equation
i) Combine the half reactions and reduce the coefficients to the lowest term when applicable.

The half reaction method is the easiest and most useful method for balancing redox equations. This is demonstrated by the following example.

Example 2.1 Balance the following reaction using the "half reaction method"

$$MnO_4^- + Cl^- + H^+ \rightarrow Mn^{2+}\ Cl_2 + H_2O$$

Solution

$$MnO_4^- \rightarrow Mn^{2+}$$

In this equation we note that there are four oxygen atoms on the left hand side of the equation. This must be balanced with 4 molecules of H_2O written on the right hand side of the equation, that is,

$$MnO_4^- \rightarrow Mn^{2+} + 4H_2O$$

The hydrogen of the water molecule on the right must be balanced with $8H^+$ written on the left side of the equation.

$$MnO_4^- + 8H^+ \rightarrow Mn^{2+} + 4H_2O$$

This equation is balanced atomically, but not electronically. The MnO_4^- has an overall chage of -1. We now proceed to determine the oxidation number for Mn in this charged compound thus. Using oxidation number of -2 for oxygen, it is seen that 8 negative charges are contributed by oxygen in this half reaction. But since the compound has an overall charge of -1 the charge on the side of oxygen is determined as $-8-(-1) = -7$. Therefore for electrical neutrality to hold, Mn must have $+7$ as its contributing charges. We further notice that Mn on the right hand side of the equation has $+2$ charges. Therefore Mn has gained 5 electron in the reaction. We then write these facts thus:

$$MnO_4^- + 8H^+ + 5e^- \rightarrow Mn^{2+} + 4H_2O \quad (a)$$

It is now seen that this equation has balanced both electronically and atomically. We then proceed to the other reactant, Cl^-

$$Cl^- \rightarrow Cl_2$$

In this equation the charge on Cl on the left side of the equation is -1 and on the right, it is 0. Therefore Cl^- has undergone oxidation reaction producing molecular chlorine. Atomically, we write the reaction as:

$$2Cl^- \rightarrow Cl_2.$$

Electronically, there are 2 negative charge on the left side of this equation which must be balanced by adding electrons on the right side.

$$2Cl^- \rightarrow Cl_2 + 2e^- \quad (b)$$

This equation is now balanced both electronically and atomically. However, there are only 2 electrons involved in the reaction as opposed to the 5 electrons encountered in the reduction of Mn. The two equations, (a) and (b), are the written thus:

$$\text{Reduction: } MnO_4^- \ 8H^+ + 5e^- \rightarrow Mn^{2+} + 4H_2O$$
$$\text{Oxidation: } 2Cl^- \rightarrow Cl_2 + 2e^-$$

Note that for the reduction, electron is written on the left hand side of the equation whereas in the oxidation equation, it is written on the right. Since the number of electrons involved in the oxidation process must equal to that in the reduction processes, we must multiply the reduction equation by 2 and the oxidation equation by 5. Doing this will result to the following equation:

$$2MnO_4^- + 16H^+ + 10e^- \rightarrow 2Mn^{2+} + 8H_2O$$
$$10Cl^- \rightarrow 5Cl_2 + 10e^-$$

Adding these two half reactions and canceling the electrons on both side of the equation, we obtain:

$$2MnO_4^- + 10Cl^- + 16H^+ \rightarrow 2Mn^{2+} + 5Cl_2 + 8H_2O$$

This is the final equation which can be seen to balance electronically and atomically.

2.3 Electrochemical System

The preceding discussion is a fundamental prerequisite for studying electron transfer in an electrochemical system. In an electrochemical system the species losing an electron is in contact with electrode at which oxidation occurs. This "lost" electron is actually transferred from the anode through the external circuit to the cathode, which serves as the electron sink. Reaction occurring at this cathode merely gains the number of electrons lost by the reaction occurring at the anode. Gaining an electron by a species during a reaction is simply saying that the species is reduced. This discussion is made clearer by the diagram in Fig. 2.1.

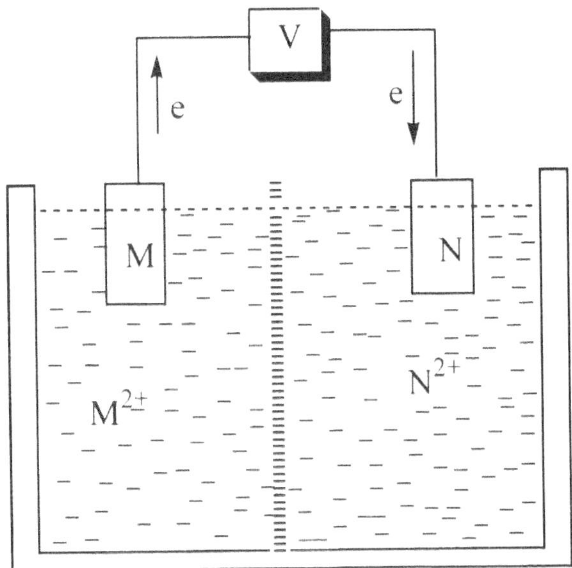

Fig. 2.1 Electron Transfer Process in Electrochemical System

In this diagram, N, the electrode on the right is serving as the electron sink of the whole system whereas the electrode on the left is the source of the electrons. The reaction occurring at each electrode may be represented thus:

Left: $M \rightarrow M^{2+} + 2e-$ (anode reaction)

Right: $N^{2+} + 2e- \rightarrow N$ (cathode reaction)

Sum: $M + N^{2+} \rightarrow M^{2+} + N$

It can be seen here that the number of electrons lost by M is equal to the number of electrons gained by N. The whole system is referred to as electrochemical cell or simply the cell.

2.4 Types of Cell

It will be recalled that in chapter one, we defined electrochemistry as the science that studies the inter-conversion of chemical and electrical energies. How can this be done? The answer lies in the electrochemical cell setup. The cells that can achieve this task may be referred to as galvanic or voltaic and electrolytic cells, respectively.

2.4.1 Galvanic Cell

When certain chemicals are mixed together, spontaneous reaction occurs. This spontaneous chemical reaction produces energy, which is dissipated as heat doing "useless" work. However, useful work can be obtained by mechanically separating the reacting species but allowing just ionic contact between them. Spontaneous reaction can still occur, but instead of heat the chemical energy is converted to electrical energy which can be used to do some useful work. A galvanic cell is an electrochemical cell that converts chemical energy arising from a chemical reaction to electrical energy.

Fig. 2.2 is a schematic diagram to illustrate a typical Galvanic or Votaic cell, so named in honor of Luigi Galvani and Alessandro Volta, who were the

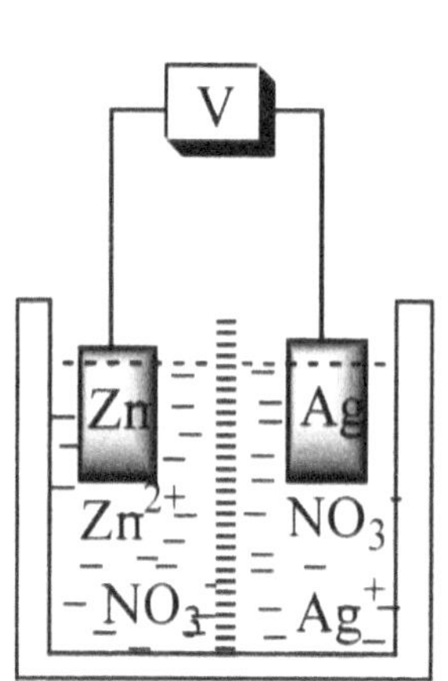

Fi.g 2.2a Galvanic Cell with Semipermeable Membrane

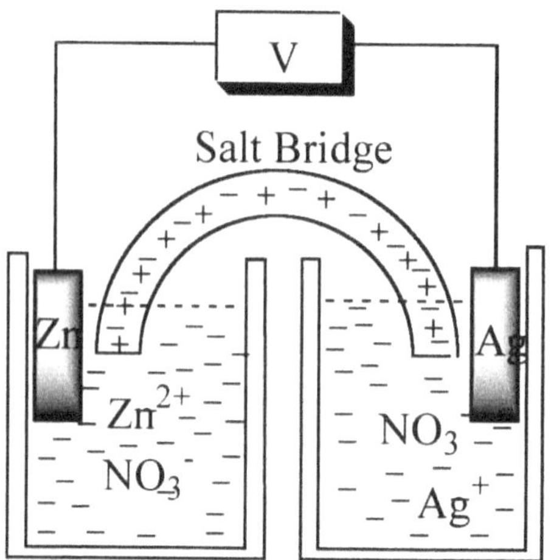

Fig. 2.2b Galvanic Cell with Salt Bridge

first scientists to discover the production of electricity through chemical change, which ultimately led to battery technology and electrochemistry (See Prologue). In the two figures, the semi-permeable membrane and the salt bridge are devices to prevent mechanical mixing of the solutions. The reaction occurring at each compartment is referred to as half reaction. They are:

At the anode: $Zn \rightarrow Zn^{2+} + 2e^-$

At the cathode: $2Ag^+ + 2e^- \rightarrow 2Ag$

Cell Reaction: $Zn + 2Ag^+ \rightarrow Zn^{2+} + 2Ag$

This reaction, as written and as set-up, will produce electrical energy, the magnitude of which is measurable by the voltmeter (V) inserted as indicated in the diagrams.

2.4.2 Electrolytic Cell

Having considered ways by which chemical energy can be converted to electrical energy, the next task is to examine how an electrical energy can be converted to a chemical energy. The type of cell that can achieve this is the electrolytic cell. In other words the use of electrical energy to carry out chemical reactions is referred to as electrolysis and the electrochemical cell set-up is called an electrolytic cell. Fig 2.3 depicts this type of cell. The set-up is for the electrolysis of molten NaCl. Upon application of electrical energy, an electric field is set-up in the solution. The Cl^- ions migrate to the positive electrode and the Na^+ ions migrate to the negative electrode.

Note that both electrodes are in contact with the same solution. The reaction occurring at the electrodes are:

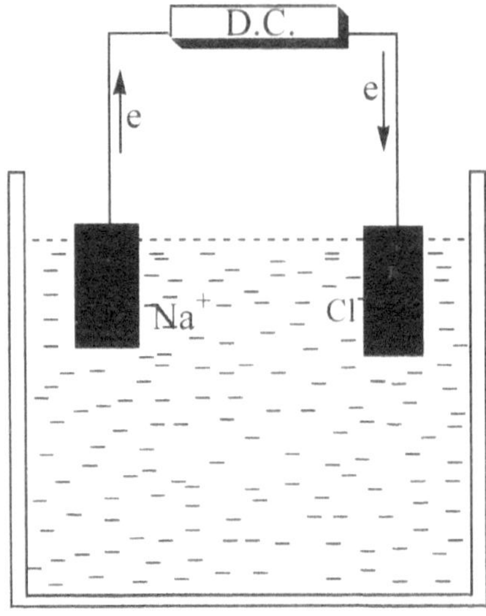

Fig. 2.3 Electrolytic Cell

At the anode: $Cl^- \rightarrow \frac{1}{2}Cl_2 + e^-$
At the Cathode: $Na^+ + e^- \rightarrow Na$

Cell Reaction: $Na^+ + Cl^- \rightarrow Na + \frac{1}{2}Cl_2$

We can remove the fractional coefficient by writing this equation as:

$$2Na^+ + 2Cl^- \rightarrow 2Na + Cl_2$$

What is the implication of this reaction? It is simply that we have, by brute force (use of external energy), forced the production of metallic sodium and chlorine gas. In the medium, they are kept separated by the electrical field. The cell reaction left to itself will prefer going to the left. To illustrate this, suppose that the direct current is suddenly switched off. A mixture of the reaction products will occur and a violent reaction between them will result. This violent reaction will, of course, release some energy. This released energy, which can now be regarded as chemical energy, will be equivalent to the amount electrical energy stored in the 2Na and Cl_2 during electrolysis. You can now appreciate how electrical energy can be converted to chemical energy by electrolysis.

2.4.3 Concentration Cell

In addition to the Galvanic (Voltaic) cell and electrolytic cell, another type of cell of import in electrochemistry is the concentration cell. While the former cells are, strictly speaking, chemical reaction cells, the latter is a cell that entails a physical transfer of ions from one phase to another. In this type of cell, the electrodes used in the half reactions are of the same kind and they are immersed in a solution of the same (common) ions. Consider a reaction in which cadmium amalgam (CdHg) is both the anode and the cathode material with the exception that they are of different concentration. (An amalgam is a metal dissolved in mercury). These electrodes are immersed in a solution of cadmium sulfate. The shorthand notation for the cell set up is as follows:

$$CdHg/Cd^{2+}/CdHg$$
$$C_1 \quad C_3 \quad C_2$$

C_1 and C_2 denote the amalgam concentration while C_3 is the cadmium concentration in the solution. We can now proceed to write the half reactions as follows:

$$CdHg(C_1) \rightarrow Cd^{2+}(C_3) + 2e^-$$
$$Cd^{2+}(C_3) + 2e^- \rightarrow CdHg(C_2).$$

The respective half cell potentials are:

$$E_1 = E^o_{1,Cd2+/Cd} - 0.059/2 \log C_1 \quad \text{anode}$$
$$E_2 = E^o_{2,Cd2+/Cd} - 0.059/2 \log C_2 \quad \text{cathode}$$

We note at this point that the E^o values for the half reactions are the same. Therefore, calculation of the cell potential using the relation $E_{cell} = E_c - E_a$ will make the E^o for the cell zero, that is

$$E_{cell} = E_2 - E_1 = E^o_{2,Cd2+/Cd} - E^o_{1,Cd2+/Cd} - 0.059/2 \log C_2 - 0.059 \log C_1$$

$$E_{cell} = E_2 - E_1 = 0 - 0.059/2 \log C_2/C_1$$
$$\therefore E_{cell} = -0.059/2 \log C_2/C_1$$

This result shows that the cell potential in a concentration cell is independent of the concentration of Cd^{2+} ions in solution. What has happened is a transfer of Cd^{2+} ions from the more concentrated amalgam to the less concentrated amalgam electrode. The pH electrode used in the determination of the H^+ concentration in a solution operates on this principle.

It will, at this point, be necessary to introduce two very important Laws that are pertinent to electrochemical reactions discussed above. These are the Ohm's and the Faraday's laws.

Ohm's Law Ohm's law states that the current, i, (amp.) flowing through a conductor is directly proportional to the electromotive force, E, (volts) impressed through the conductor and inversely proportional to the resistance of the conductor, R, (ohms). Mathematically this can be expressed as:

$$E = iR \text{ or } i = E/R \tag{2.1.}$$

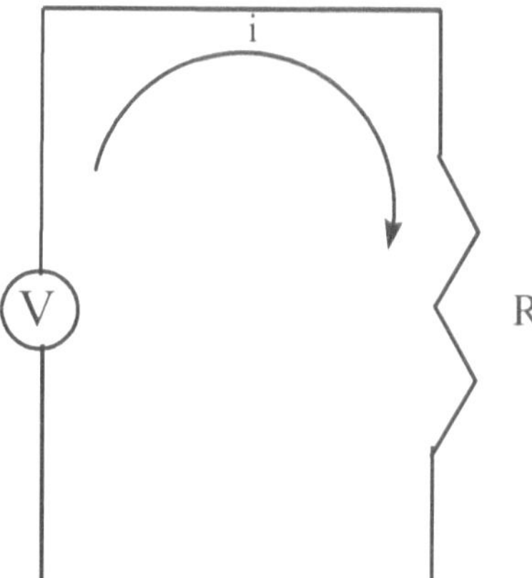

If various values of V (Potential) are plotted with the associated resistor values the resultant slope will be equal to the current in the circuit as can be seen the following data and the corresponding plot.

Potential, E   Resistance, R (Ohms)

9	18
24	48
30	60
50	100
60	120
100	200
110	220

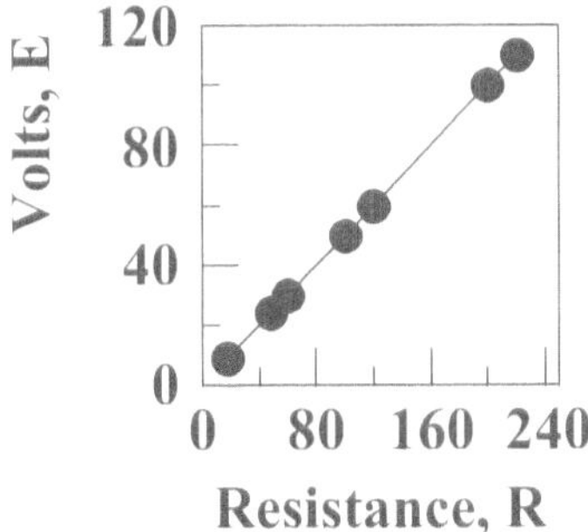

The value of the slope of this plot is 0.5 Volt/Ohm, which is 0.5 Ampere. This value corresponds to the current that flows in the simple circuit shown above if the voltage source and the resistor value vary.

Faraday's First Law This law states that the quantity of chemical substance produced or consumed at an electrode is proportional to the quantity of electricity passed through the cell. The unit of this quantity is the Coulomb, C. There are approximately 96,500 C. in one Faraday, F, which is equal to the passage of one mole of electrons. This law can be expressed mathematically as:

$$Q = it \qquad\qquad 2.2$$

where i = current in amps, and t is time in seconds.

The quantitative relationship between the applied voltage and quantity of material produced in an electrolytic reaction is therefore

$$m = \frac{QM}{nF} = 1.0363 \times 10^{-5} \frac{QM}{n}$$

In this equation F is the Faraday constant and M and n are the molar mass and number of electrons, respectively. It will be noted that the value of n corresponds to the valence of the ion in the solution.

Faraday's Second Law When equal amount of electricity or charge is passed in electrolysis solution of different substances, the amount of materials produced at the electrodes is proportional to that substance equivalent weight. The equivalent weight of a given substance is the molar mass of the substance divided by the number of electrons (n) that were involved in the reaction. Some examples of this will suffice to make this clearer.

$$\text{Equivalent Weight of } Mg^{2+} = \frac{\text{Molar Mass of Mg}}{2}$$

$$\text{Equivalent Weight of } Na^{+} = \frac{\text{Molar Mass of Na}}{1}$$

$$\text{Equvalent Weight of } Al^{3+} = \frac{\text{Molar Mass of Al}}{3}$$

$$\text{Equivalen Weight of } CO_3^{2-} = \frac{\text{Molar Mass of } CO_3}{2}$$

Problem 2.1

In a given electrolytic reaction, a voltage of 1.5V was passed in a NaCl solution of 0.50 Ohms resistance for 10 minutes. Calculate the weight of chlorine gas produced.

Solution

Since it is necessary first to find the total current passed in the 10 minutes interval, we shall use the Ohm's law to determine this quantity, thus:

$$i = E/R = 1.5V/0.5 \text{ Ohm} = 3 \text{ amps.}$$

Now we apply the Faraday's law to find the quantity of electricity passed.

$Q = it = 3$ amps x 10 minutes x 60 secs/min = 1800 Coul.

We then find the number of Faradays in this quantity of electricity used.

96,500 Coul = 1F
1,800 Coul = 1F/96,500 Coul x 1,800 Coul = 0.0186F

Notice the reaction taking place

$2Cl^- \rightarrow Cl_2 + 2e^-$

This reaction says that to produce 1 mole of Cl_2 (Molecular Mass = 70.90) 2 electrons or 2F must be used. But we have only used 0.0186F. The question is how many grams of Cl_2 is produced by this number of Faraday.

2F will produce 70.90g Cl_2
$\therefore$ 0.0186F will produce 70.90g Cl_2/2F x 0.0186F = 0.659g Cl_2

2.5 Potential of Electrochemical Cell

In the previous sections we discussed the inter-converstion of electrical and chemical energies with a transfer of a unit of charge across an electrochemical cell. Energy is the ability to do work. The unit of energy of interest here is the volt.

The operational definition of volt is joule per charge. This is represented mathematically as:

$$1V = \frac{1J}{C} \quad \text{where } C = \text{Coul (quantity of charge).} \quad 2.3$$

In an electrochemical cell, there are two electrodes. For an electron transfer to take place between these electrodes, there must be a potential difference existing between them. Some work must be done to transfer a charge from one electrode to the other. The amount of work done in transferring one unit of charge across these electrodes is measured in volts and is a measure of the cell potential. Put another way, cell potential is a measure of the potential energy difference between two electrodes of a cell.

2.5.1 Electrode Potential

From the definition of cell potential it follows that there exists a potential at any given electrode. This means that for any electrode in contact with a solution (electrolyte) there must be a potential development at the electrode-electrolyte interface. But how do we determine this? It is impossible to measure the potential of a single electrode. To overcome this dilemma a reference system becomes a necessity.

In most electrochemical measurements, the reference system (electrode) devised is the hydrogen electrode which is arbitrarily assigned a potential of zero.

A Standard Hydrogen Electrode (SHE) consists of a cell with an "inert" metal electrode usually platinum black dipping into hydrochloric acid solution of 1.0 M concentration. The platinum is bubbled with hydrogen gas at a pressure of 1.0 atm.

The reaction taking place at this electrode is as follows:

$$H_2(g) \rightarrow 2H^+ + 2e^-$$

The cell diagram containing a SHE is shown in Fig. 2.4.

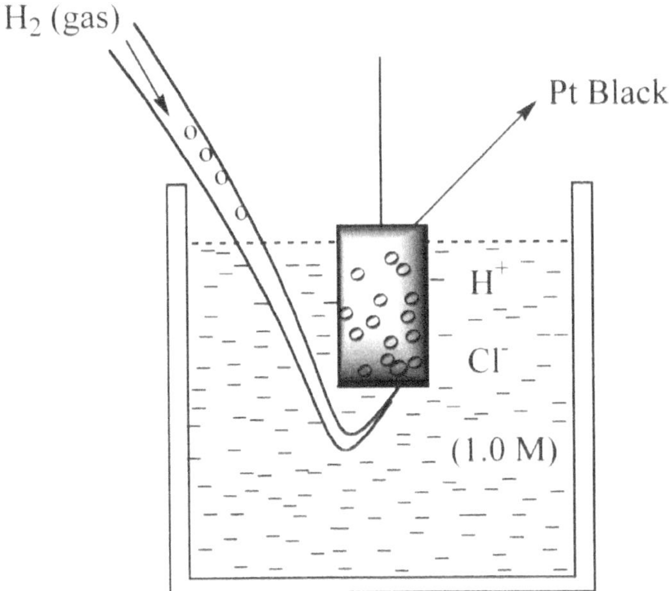

Fig. 2.4 Standard Hydrogen Electrode

Since the SHE has arbitrarily been assigned a potential of zero, it only means that we can couple it with any electrode of our interest and measure the developed potential of the cell. The potential so obtained is, of course, the potential of the electrode of interest. Although SHE is the reference electrode most often referred to, it is inconvenient to work with. As a result other more convenient reference electrodes are routinely used in electrochemical measurements. Most prominent among them are the silver/silver chloride (Ag/AgCl) and the calomel (Hg_2Cl_2), usually referred to as SCE, electrodes. The following table gives the potential of the most commonly used reference electrodes. These potentials are compared with the hydrogen and calomel reference electrode potentials.

Table 2.1 Potentials (V) of Reference Electrodes at 25 °C‡

	versus	
	NHE*	SCE#
NHE	0.0000	-0.2412
Ag/AgCl, sat'd KCl	0.1970	-0.0450
Hg/Hg$_2$Cl$_2$, SCE, sat'd NaCl	0.2360	
Hg/Ag$_2$Cl$_2$, SCE, sat'd KCl	0.2412	0.0000
Hg/Hg$_2$Cl$_2$, NCE, KCl(1.0 M)	0.2801	

Hg/Hg$_2$Cl$_2$, KCl(0.10 M)	0.3337	
Hg/Hg$_2$SO$_4$, sat'd K$_2$SO$_4$	0.6400	0.4000
Hg/Hg$_2$SO$_4$, H$_2$SO$_4$(0.50M)	0.6800	
Hg/HgO, NaOH(0.10 M)	0.9260	0.6850

‡ - Values taken from *Electrochemical Methods Fundamentals and Applications* by Allen J. Bard and Larry R. Faulkner, Wiley, NY 1980

*- Normal Hydrogen Electrode; # - Saturated Calomel Electrode

2.6 E^o Values and EMF

Electromotive force (EMF) is the value in volts of the potential developed in a galvanic cell. If this value is obtained at standard conditions, that is, when the reactants and products of the system are at unit activity or concentration and the temperature is 25 ^{o}C at 1.0 atm., it is referred to as E^o (called E-zero). The E^o values of many electrode materials have been determined and some of the values are listed in Table 2.2.

Table 2.2 E^0 Values of Some Half Reactions

<u>Reactions</u>	<u>E^0, Volts</u>
$Sr^+ + e^- \rightleftarrows Sr$	-4.10
$Ra^+ + e^- \rightleftarrows Ra$	-4.0
$3N_2 + 2H^+ \rightleftarrows 2NH_3$	-3.40
$Eu^{3+} + 3e^- \rightleftarrows Eu$	-3.395
$Sm^{2+} + 2e^- \rightleftarrows Sm$	-3.121
$Li^+ + e^- \rightleftarrows Li$	-3.0401
$Rb^+ + e^- \rightleftarrows Rb$	-2.98
$K^+ + e^- \rightleftarrows K$	-2.931
$Cs^+ + e^- \rightleftarrows Cs$	-2.923
$Ra^{2+} + 2e^- \rightleftarrows Ra$	-2.916
$Ba^{2+} + 2e^- \rightleftarrows Ba$	-2.906
$Sr^{2+} + 2e^- \rightleftarrows Sr$	-2.892
$Ca^{2+} + 2e^- \rightleftarrows Ca$	-2.869
$Yb^{2+} + 2e^- \rightleftarrows Yb$	-2.797
$Na^+ + e^- \rightleftarrows Na$	-2.71
$Mg^{2+}\ 2e^- \rightleftarrows Mg$	-2.70
$Cm^{3+}\ 3e^- \rightleftarrows Cm$	-2.70
$Mg(OH)_2 + 2e^- \rightleftarrows Mg + 2OH^-$	-2.69
$BeO + H_2O + 2e^- \rightleftarrows Be + 2OH^-$	-2.613
$Ac^{3+} + 3e^- \rightleftarrows Ac$	-2.6
$La^{3+} + 3e- \rightleftarrows La$	-2.522
$No^{3+} + 3e^- \rightleftarrows No$	-2.5
$Ce^{4+} + e^- \rightleftarrows Ce^{3+}$	-2.483
$Pr^{3+} + 3e^- \rightleftarrows Pr$	-2.462
$Nd^{3+} + 3e^- \rightleftarrows Nd$	-2.431
$Pm^{3+} + 3e^- \rightleftarrows Pm$	-2.423

$Sm^{3+} + 3e^- \rightleftarrows Sm$	-2.414
$Eu^{3+} + 3e^- \rightleftarrows Eu$	-2.409
$Bk^{3+} + 3e^- \rightleftarrows Bk$	-2.4
$Gd^{3+} + 3e^- \rightleftarrows Gd$	-2.397
$Tb^{3+} \ 3e^- \rightleftarrows Tb$	-2.391
$Am^{3+} + 3e^- \rightleftarrows Am$	-2.38
$Y^{3+} + 3e^- \rightleftarrows Y$	-2.372
$Mg^{2+} + 2e^- \rightleftarrows Mg$	-2.363
$Dy^{3+} + 3e^- \rightleftarrows Dy$	-2.353
$Ho^{3+} + 3e^- \rightleftarrows Ho$	-2.319
$Er^{3+} + 3e^- \rightleftarrows Er$	-2.296
$Tm^{3+} + 3e^- \rightleftarrows Tm$	-2.278
$Yb^{3+} + 3e^- \rightleftarrows Yb$	-2.267
$Lu^{3+} + 3e^- \rightleftarrows Lu$	-2.255
$Cf^{3+} + 3e^- \rightleftarrows Cf$	-2.1
$Fm^{3+} + 3e^- \rightleftarrows Fm$	-2.1
$Sc^{3+} \ 3e^- \rightleftarrows Sc$	-2.07
$Pu^{3+} \ 3e^- \rightleftarrows Pu$	-2.031
$Pa^{3+} + 3e^- \rightleftarrows Pa$	-1.95
$Th^{4+} + 4e^- \rightleftarrows Th$	-1.899
$Np^{3+} + 3e^- \rightleftarrows Np$	-1.856
$Be^{2+} + 2e^- \rightleftarrows Be$	-1.847
$U3+ + 3e- \rightleftarrows U$	-1.798
$BeO + 2H^+ + 2e^- \rightleftarrows Be + H_2O$	-1.785
$Hf^{4+} + 4e^- \rightleftarrows Hf$	-1.700
$Pa^{4+} + 4e^- \rightleftarrows Pa$	-1.7
$Al^{3+} \ 3e^- \rightleftarrows Al$	-1.662
$Ti^{2+} + 2e^- \rightleftarrows Ti$	-1.630

$Ce^{4+} + e^- \rightleftarrows Ce^{3+}$	-1.61
$Zr^{4+} + 4e^- \rightleftarrows Zr$	-1.539
$Mn^{3+} + 3e^- \rightleftarrows Mn$	-1.185
$Mn^{2+} + 2e^- \rightleftarrows Mn$	-1.180
$V^{2+} + 2e^- \rightleftarrows V$	-1.175
$Te + 2e^- \rightleftarrows Te^{2-}$	-1.143
$Nb^{3+} + 3e^- \rightleftarrows Nb$	-1.099
$WO_4^{2-} + 4H_2O + 6e^- \rightleftarrows W + 8OH^-$	-1.05
$Se + 2e^- \rightleftarrows Se^{2-}$	-0.924
$Cr^{2+} + 2e^- \rightleftarrows Cr$	-0.913
$SiO_2 + 4H^+ + 4e^- \rightleftarrows Si + 2H_2O$	-0.840
$Zn^{2+} + 2e^- \rightleftarrows Zn$	-0.7618
$Ta_2O_5 + 10H^+ \ 10e^- \rightleftarrows 2Ta + 5H_2O$	-0.750
$Cr^{3+} + 3e^- \rightleftarrows Cr$	-0.744
$Ga^{3+} + e^- \rightleftarrows Ga^{2+}$	-0.677
$SbO_2^- + 2H_2O + 3e^- \rightleftarrows Sb + 4OH^-$	-0.67
$As + 3H^+ + 3e^- \rightleftarrows AsH_3$	-0.608
$ReO_2 + 2H_2O + 3e^- \rightleftarrows Re + 4OH^-$	-0.577
$Ga^{3+} + 3e^- \rightleftarrows Ga$	-0.529
$S + 2e^- \rightleftarrows S^{2-}$	-0.51
$Sb + 3H^+ \ 3e^- \rightleftarrows SbH_3$	-0.510
$Fe^{2+} + 2e^- \rightleftarrows Fe$	-0.447
$In^{3+} + 2e^- \rightleftarrows In^+$	-0.443
$Cr^{3+} + e^- \rightleftarrows Cr^{2+}$	-0.407
$Cd^{2+} + 2e^- \rightleftarrows Cd$	-0.4030
$H_2PO_2^- + 2H^+ + e^- \rightleftarrows P + H_2O$	-0.391
$Ti^{3+} + e^- \rightleftarrows Ti^{2+}$	-0.368
$In^{3+} + 3e^- \rightleftarrows In$	-0.3382

$Tl^+ + e^- \rightleftharpoons Tl$	-0.336
$H_3PO_4 + H^+ + 2e^- \rightleftharpoons H_2PO_3^- + H_2O$	-0.329
$Mn^{3+} + 3e^- \rightleftharpoons Mn$	-0.283
$Co^{2+} + 2e^- \rightleftharpoons Co$	-0.28
$Ni^{2+} + 2e^- \rightleftharpoons Ni$	-0.257
$Mo^{3+} + 3e^- \rightleftharpoons Mo$	-0.20
$Sn^{2+} + 2e^- \rightleftharpoons Sn$	-0.151
$Pb^{2+} + 2e^- \rightleftharpoons Pb$	-0.126
$WO_2 + 4H^+ + 4e^- \rightleftharpoons W + 2H_2O$	-0.119
$WO_3 + 6H^+ \, 6e^- \rightleftharpoons W + 3H_2O$	-0.090
$Ti^{4+} + e^- \rightleftharpoons Ti^{3+}$	-0.04
$Fe^{3+} + 3e^- \rightleftharpoons Fe$	-0.037
$D^+ + e^- \rightleftharpoons 1/2D_2$	-0.0034
$H^+ + e^- \rightleftharpoons 1/2H_2$	0.000
$Ge^{2+} + 2e^- \rightleftharpoons Ge$	+0.01
$Cu^{2+} + e^- \rightleftharpoons Cu^+$	+0.153
$Sn^{4+} + 2e^- \rightleftharpoons Sn^{2+}$	+0.154
$Mn^{4+} + 4e^- \rightleftharpoons Mn$	+0.195
$At_2 + 2e^- \rightleftharpoons 2At$	+0.2
$Bi^{3+} + 3e- \rightleftharpoons Bi$	+0.200
$CO_2 + 4H^+ + 4e^- \rightleftharpoons C + 2H_2O$	+0.207
$AgCl + e^- \rightleftharpoons Ag + Cl^-$	+0.222
$Re_2O + 4H^+ + 4e^- \rightleftharpoons Re + 2H_2O$	+0.2519
$Cu^{2+} + 2e^- \rightleftharpoons Cu$	+0.3419
$Fe(CN)_6^{3-} + e^- \rightleftharpoons Fe(CN)_6^{4-}$	+0.358
$Tc^{2+} + 2e^- \rightleftharpoons Tc$	+0.400
$\frac{1}{2}O_2 + H_2O + 2e^- \rightleftharpoons 2OH^-$	+0.401
$Ru2+ + 2e- \rightleftharpoons Ru$	+0.455

$$CO + 2H^+ + 2e^- \rightleftharpoons C + H_2O \qquad +0.518$$

$$Cu^+ + e^- \rightleftharpoons Cu \qquad +0.521$$

$$I_2 + 2e^- \rightleftharpoons 2I^- \qquad +0.5355$$

$$Te^{4+} + 4e^- \rightleftharpoons Te \qquad +0.56$$

$$MnO_4^- + e^- \rightleftharpoons MnO_4^{2-} \qquad +0.564$$

$$Rh^{2+} + 2e^- \rightleftharpoons Rh \qquad +0.60$$

$$Po^{2+} + 2e^- \rightleftharpoons Po \qquad +0.651$$

$$Fe^{3+} + e^- \rightleftharpoons Fe^{2+} \qquad +0.771$$

$$Hg_2^{2+} + 2e^- \rightleftharpoons 2Hg \qquad +0.7973$$

$$Ag^+ + e^- \rightleftharpoons Ag \qquad +0.800$$

$$BO + 2H^+ \; 2e^- \rightleftharpoons B + H_2O \qquad +0.806$$

$$OsO_4 + 8H^+ \; 8e^- \rightleftharpoons Os + 4H_2O \qquad +0.85$$

$$Pd^{2+} + 2e^- \rightleftharpoons Pd \qquad +0.951$$

$$OsO_4^{2-} + 8H^+ + 6e^- \rightleftharpoons Os + 4H_2O \qquad +0.994$$

$$Br_2 + 2e^- \rightleftharpoons 2Br^- \qquad +1.066$$

$$CCl_4 + 4e^- \rightleftharpoons C + 4Cl \qquad +1.18$$

$$Pt^{2+} + 2e^- \rightleftharpoons Pt \qquad +1.88$$

$$MnO_2 + 4H^+ + 2e^- \rightleftharpoons Mn^{2+} \; 2H_2O \qquad +1.224$$

$$Tl^{3+} + 2e^- \rightleftharpoons Tl^+ \qquad +1.25$$

$$Cr_2O_7^{2-} + 14H^+ \; 6e^- \rightleftharpoons 2Cr^{3+} + 7H_2O \qquad +1.33$$

$$Cl_2 + 2e^- \rightleftharpoons 2Cl^- \qquad +1.395$$

$$PbO_2 + 4H^+ + 2e^- \rightleftharpoons Pb^{2+} + 2H_2O \qquad +1.455$$

$$Au^{3+} + 3e^- \rightleftharpoons Au \qquad +1.498$$

$$MnO_4^- + 4H^+ + 3e^- \rightleftharpoons MnO_2 + 2H_2O \qquad +1.695$$

$$Ce^{4+} + e^- \rightleftharpoons Ce^{3+}(1F/HCO_4) \qquad +1.70$$

$$Co^{3+} + e^- \rightleftharpoons Co^{2+} \qquad +1.84$$

$$OH + e^- \rightleftharpoons OH^- \qquad +2.02$$

$$F_2 + e^- \rightleftharpoons 2F^- \qquad +2.866$$

Source: Data were taken from the Tables of Standard Electrode Potentials by G. Milazzo and S. Caroli (1978), John Wiley, New York.

2.6.1 E^o Values and their consequences

(a) (i) Half Reaction.
The listed E^o values are for the half reactions of a given cell. A half reaction is the reaction taking place at an electrode, for example.

$$Sn^{4+} + 2e^- \rightarrow Sn^{2+} \qquad E^o = +0.154V$$

By convention the reaction is written as reduction and the E^o value of the half reaction is always included.

(ii) When two half reactions are coupled the one with greater negative E^o value is reversed and its E^o value is also reversed. For example:

$$Zn^{2+} + 2e^- \rightarrow Zn \qquad E^o = -0.763 \text{ V}$$
$$Cu^{2+} + 2e^- \rightarrow Cu \qquad E^o = +0.340 \text{ V}$$

When these two half reactions are coupled the following is obtained.

$$Zn + Cu^{2+} \rightarrow Zn^{2+} + Cu$$

$$E^o = +0.340 -(-0.763) \text{ V} = (+0.340 + 0.763) \text{ V} = +1.103 \text{ V}$$

(b) Magnitude of E^o

The more positive the value of E^o the greater the tendency for the reaction to go from left to right, that is, the greater its oxidizing ability. Conversely, the more negative the E^o the smaller the tendency the reaction has to proceed from left to right. In this case it will have greater reducing ability. Also it will have greater tendency to reverse.

(c) The Coefficient of Half Reactions

The coefficients in half reactions do not affect the E^o values attached to them. For example:

$$\tfrac{1}{2}F_2(g) + e^- \rightarrow F^-$$
$$F_2(g) + 2e^- \rightarrow 2F^-$$
$$2F_2(g) + 4e^- \rightarrow 4F^-$$

The E^o value for each of these fluorine molecules is +2.87V.

2.7 Cell Reaction and Potential

We have earlier determined that a cell potential is the potential difference between the half reactions of a cell. Half reactions have also been discussed. The cell reaction, therefore, is an addition of the half reactions. In addition it will be noted that if the cell is to operate as a galvanic cell, the obtained E^o of the cell must be positive. This follows from a thermodynamic relationship of Gibbs Free Energy with cell potential.

$$\Delta G = -nFE \qquad\qquad 2.4a$$
$$\Delta G^o = -nFE^o \qquad\qquad 2.4b$$

Thermodynamically, a spontaneous reaction implies a negative value of ΔG. Since galvanic cells operate spontaneously, the obtained E or E^o must be positive. In an electrochemical reaction how do we achieve this? The following reaction will explain the steps necessary to achieve our objective of obtaining a positive E or E^o.

Consider the following half reactions, all written as reduction:

$$Ni^{2+} + 2e^- \rightarrow Ni \qquad\qquad E^o = -0.23V$$
$$2Fe^{+3} + 2e^- \rightarrow 2Fe^{+2} \qquad\qquad E^o = +0.77V$$

Evaluation of the E^o values of these half reactions and bearing in mind our discussion in the last section that the reaction with more negative value will have a tendency to reverse, will make us want to reverse the Ni^{+2} half reaction. If we do as our intuition directs and do it right we will obtain the following equation:

$$Ni \rightarrow Ni^{+2} + 2e^- \qquad\qquad E^o = 0.23V$$

$$2Fe^{+3} + 2e^- \rightarrow 2Fe^{+2} \qquad\qquad E^o = 0.77V$$

Cell reaction: $Ni + 2Fe^{+3} \rightarrow Ni^{+2} + 2Fe^{+2}$ E^o cell $= 1.00V$

What does this last equation tells us? It tells us three things:

1.   It tells us that Ni in this cell reaction is oxidized and therefore makes up the anode half reaction while the Fe^{+3} makes up the cathode half reaction.

2.   It tells us that it is the anode half reaction that is reversed together with its E^o value.

This brings us to a very important conclusion. That is, the E^o_{cell} or E_{cell} of any electrochemical reaction operating galvanically, can be determined by the following relation:

$$E_{cell} = E_{cathode} - E_{anode} \qquad\qquad 2.5$$

3.   It tells us that the reaction as written will proceed spontaneously. Had the E^o cell been negative the reaction as written will not proceed spontaneously.

2.8 Shorthand Notation for Cell Representation

Consider the Ni-Fe cell reaction discussed in the previous section. Since this reaction is operating in a galvanic mode, it is obvious that the half reactions must be mechanically separated. The ionic contact that is required is accomplished either by a salt bridge or by a semi-permeable membrane, see Fig. 2.2. Let us examine the anode compartment.

The electrode is most assuredly a nickel plate or wire. The solution with which this electrode is in contact is $Ni(SO_4)_2$. At the cathode, the electrode is an iron bar or wire in contact with a solution of $Fe_2(SO_4)_3$. Both the $Ni(SO_4)_2$ and $Fe_2(SO_4)_3$ have concentrations x and y respectively, and are separated as described above.

An electrode in solution, in contact with its ions, is indicated by a slash or a bar. This slash or bar means an interface. If a salt-bridge or membrane is separating the solution, a double slash is used. To sum it up, the following notation is used for the Ni-Fe cell.

$$\mathbf{Ni(s) \mid Ni^{2+}(xM) \mid \mid Fe^{3+}(yM) \mid Fe(s)}$$

Note that the anode is written at the extreme left while the cathode is written at the extreme right. In other words, the short hand notation for a cell reads:

Anode-solution-salt bridge   or   membrane-solution-cathode   or anode/solution//solution /cathode.

2.9 Electrode Potential and Equilibrium constant from Thermodynamic Consideration

Free energy changes occurring in a reaction system at standard conditions (STP) is usually estimated from the thermodynamic parameter, ΔG^o. However, most often reactions are carried out at conditions different from STP conditions. It is therefore necessary to estimate the free energy of reactions for such systems. The generalized Gibbs free energy change for a reaction system shown in equation 2.6 is used.

$$\Delta G = \Delta G^o + RT\ln K \qquad\qquad 2.6$$

It can be seen from this equation that for systems at equilibrium, whose reactants and products are at unit activity or concentration, the resulting free energy change is equal to the free energy change at STP. Furthermore, ΔG for the system at equilibrium is zero. At this condition, equation 2.6 reduces to equation 2.7, consistent with thermodynamic considerations.

$$\Delta G^o = -RT\ln K \qquad\qquad 2.7$$

We have earlier shown the relationship between ΔG and E from electrochemical reactions, that is,

$$\Delta G = -nFE \quad \text{and} \quad \Delta G^o = -nFE^o$$

Inserting these equations into equation 2.6 we obtain equation 2.8.

$$nFE = -nFE^o + RT\ln K \qquad\qquad 2.8$$

This equation can be rewritten as

$$nF(E^\circ - E) = RT\ln K \qquad\qquad 2.9$$

Rearrangement of this equation and solving for E results to equation 2.10.

$$E = E^\circ - \frac{RT}{nF}\ln K \qquad\qquad 2.10$$

In the above equations

    R = universal gas constant, 8.314 J/K-mole

    T = Absolute temperature, K

    K = equilibrium constant

    n = number of electrons; F = Faraday's constant, 96,500 Coul.

If we put all the constants and converting natural logarithm to logarithm base 10, we will obtain the following equation:

$$E = \frac{1.98 \times 10^{-4}T}{n}\log K \qquad\qquad \mathbf{2.11}$$

The value, that is, the pre-logarithmic factor of equation 2.11 at different temperatures and values of n are listed in Table 2.3.

Table 2.3 Values of 2.303RT/nF in Volts at Different Temperatures and Different values of n

T,K	n = 1	n = 2	n = 3
273	0.0541	0.0272	0.0181
275	0.0552	0.0276	0.0184
283	0.0562	0.0281	0.0187
288	0.0571	0.0286	0.0190
293	0.0580	0.0290	0.0194
298	0.0592	0.0295	0.0197
303	0.0601	0.0299	0.0200
308	0.0610	0.0306	0.0204
313	0.0622	0.0311	0.0207
315	0.0631	0.0316	0.0210
323	0.0640	0.0320	0.0214
328	0.0652	0.0325	0.0217
333	0.0661	0.0329	0.0220

The implication of equation 2.11 as illustrated in Table 2.3 is that the electrode potential is a function of temperature, the number of exchanged electrons in the reaction and the equilibrium concentration of the reacting substances.

At $25^{\circ}C$, equation 2.11 becomes equation 2.12.

$$E = \frac{0.059}{n} \log K \qquad\qquad 2.12$$

Consider the following chemical reaction at equilibrium:

$$aA + bB \; \rightleftarrows \; cC + dD$$

If the activity coefficient is neglected, the equilibrium expression for the above reaction will be:

$$K = \frac{[C]^{c}[D]^{d}}{[A]^{a}[B]^{b}} \qquad\qquad 2.13$$

If equations 2.12 and 2.13 are combined, equation 2.14 results.

$$E = \frac{0.059}{n} \log \frac{[C]^{c}[D]^{d}}{[A]^{a}[B]^{b}} \qquad\qquad 2.14$$

If the reacting and the resulting substances are at unit concentrations, and the number of electrons, n, transferred during the reaction process is 1, then the E that will be obtained will, of course, correspond to our earlier defined E^{o}, that is:

$$E^{o} = 0.059 \log K \text{ at standard conditions}$$

Suppose that the reaction was conducted at temperatures other than $25^{\circ}C$ and the reacting and resulting substances are at concentrations other than unity, the resulting E will certainly differ from the E obtained at standard conditions, and we may represent this new E in relation with the E^{o} thus:

$$E = E^{o} - \frac{RT}{nF} \ln K \qquad\qquad 2.15$$

This is the well known, and inevitable, Nernst equation. It simply states that the electrochemical potential of any reacting substances in equilibrium is a function of the standard emf (E^o) and the ratio of the concentrations of the reacting and resulting substances.

Equation 2.13 is of fundamental importance and, I may add, a "primus inter pares" of all electrochemical equations in their various forms.

It is in order, at this point, to solve some problems to illustrate the significance of this equation.

Problem 2.2

Consider the following reaction taking place at 25 °C:

$$Fe^{2+} + MnO_4^- \rightarrow Fe^{3+} + Mn^{2+}$$
$$E^o_{Fe3+/Fe2+} = +0.771V$$
$$E^o_{MnO4-/Mn2+} = +1.510V$$

Suppose at equilibrium the substances have the following concentrations:

$$[Fe^{2+}] = 0.01M, [Fe^{3+}] = 0.05M,$$
$$[MnO_4^-] = 0.001M, [Mn^{2+}] = 0.10M, [H^+] = 0.1M$$

Calculate the resulting potential. As written, is the cell galvanic or electrolytic?

To solve this problem the equation has to be balanced.

$$Fe^{2+} \rightarrow Fe^{3+} + e^- \qquad\qquad (1)$$
$$5e^- + 8H^+ + MnO_4^- \rightarrow Mn^{2+} + 4H_2O \qquad (2)$$

Multiplication of equation (1) by 5 and (2) by 1 to balance the charges, and summing up gives:

$$5Fe^{2+} + MnO_4^- + 8H^+ \rightarrow 5Fe^{3+} + Mn^{2+} + 4H_2O \text{ (balanced)}$$

Application of the Nernst equation will give

$$E = E^o - \frac{0.059 \text{ V log } (0.05)^5 (0.10)}{5 \ (0.01)^5 (0.001) (0.1)^8}$$

To obtain the E^o we evaluate the anode material and the cathode material.

The balanced equation tells us that the Fe is the anode material since it goes from +2 to +3 and, of course, the Mn is the cathode material.

So the E^o_{cell} of the cell reaction is $E^o_{cathode} - E^o_{anode}$

$$= (1.510V\text{-}0.771V) = 0.739V$$
$$\therefore \ E = 0.739V - \frac{0.059 \text{ V}}{5} (\log 3.13x10^{-9} - \log 1.0x10^{-21})$$
$$= (0.739 - 0.147) \text{ V} = 0.529 \text{ V}$$

As written the cell is galvanic since the obtained E is positive.

Example 2.3   Using the solution in problem 2.2 calculate the equilibrium constant and ΔG for the reaction between MnO_4^- and Fe^{2+}.

Solution Since the cell potential of this reaction is known, we can use the equation $\Delta G = -nFE = -RT\ln K$. First we calculate ΔG.

$$n = 5 \text{ or simply 5 equivalents, eq; } F = 96500 \text{ Coul;}$$
$$E = 0.529 \text{ V}$$

a) Therefore $\Delta G = -5 \text{ eq x } 96500 \text{ Coul/eq x } 0.529 \text{ V}$
$$= -255.2 \text{ x } 10^3 \text{ C-V.}$$

Since $1V = 1J/C$, C-V = J. The solution therefore is:
$$\Delta G = -255.2 \text{ x } 10^3 J \text{ or } -255.2 \text{ kJ.}$$

b) We can use the above given relation: $E = RT /nF \ln K$
   $\ln K = nFE/RT$;
   $$\log K = \frac{(5 \text{ eq x } 96500 \text{ C/eq x } 0.529 \text{ V})}{2.303 \text{ x } 8.314 \text{ J/K.mol x } 298 \text{ K}}$$
   $\log K = 44.7335; \ K = 5.4 \text{ x } 10^{44}$

Most of our discussions so far have concentrated on reaction carried out at standard conditions, which imply a temperature of 25°C. But from the foregone discussions, we have shown that electrode potentials are a function of temperature. It is quite possible, therefore, to determine standard electrode potentials at temperatures other than 25°C. From thermodynamic considerations, the temperature coefficient, of the standard electrode potentials, have been determined. The generalized equation, electrode potential temperature coefficient, for use in such determination is given in equation 2.16.

$$E^{\circ} = E^{\circ}_{25} + dE/dT(T-25) \qquad\qquad 2.16$$

This equation states that a knowledge of a given standard electrode potential and its temperature coefficient, dE/dT, could be used to calculate the standard potential of such electrode at any other temperature, T. Values of the temperature coefficient of some electrode materials are given on Table 2.4

Table 2.4 Temperature Coefficient of Selected Electrode Reactions

Electrode Reaction	dE/dT, mV/K*
$Na^+ + e^- \rightarrow Na$	-0.772
$Mg^{2+} + 2e^- \rightarrow Mg$	+0.103
$Al^{3+} + 3e^- \rightarrow Al$	+0.504
$Zn^{2+} + 2e^- \rightarrow Zn$	+0.091
$Fe^{2+} + 2e^- \rightarrow$	+0.052
$Cd^{2+} + 2e^- \rightarrow Cd$	-0.093
$Sn^{2+}\ 2e^- \rightarrow Sn$	-0.282
$Pb^{2+} + 2e^- \rightarrow Pb$	-0.451
$2H^+ + 2e^- \rightarrow H_2(SHE)$	0.000
$Cu^{2+} + e^- \rightarrow Cu^+$	+0.073
$AgCl + e^- \rightarrow Ag + Cl^-$	-0.653
$Hg_2Cl_2 + 2e^- \rightarrow 2Hg + 2Cl^-$	-0.317
$Cu^{2+} + 2e^- \rightarrow Cu$	+0.008
$Cu^+ + e^- \rightarrow Cu$	-0.058
$I_2 + 2e^- \rightarrow 2I^-$	-0.148
$Hg_2SO_4 + 2e^- \rightarrow 2Hg + SO_4^{2-}$	-0.826
$Fe^{3+} + e^- \rightarrow Fe^{2+}$	+1.188
$Ag+ + e- \rightarrow Ag$	-1.000
$Br_2(aq) + 2e^- \rightarrow 2Br^-$	-0.478
$O_2 + 4H^+ + 4e- \rightarrow 2H_2O$	-0.846
$MnO_2 + 4H^+ + 2e^- \rightarrow Mn^{2+} + 2H_2O$	-0.661
$Cl_2 + 2e^- \rightarrow 2Cl^-$	-1.260
$Mg(OH)_2 + 2e^- \rightarrow Mg + 2OH^-$	-0.074
$Al(OH)_3 + 3e^- \rightarrow Al + 3OH^-$	-0.06
$Cr(OH)_3 + 3e^- \rightarrow Cr + 3OH^-$	-0.11
$Zn(OH)_2 + 2e^- \rightarrow Zn + 2OH^-$	-0.131
$ZnCO_3 + 2e^- \rightarrow Zn + CO_3^{2-}$	-0.293
$Fe(OH)_2 + 2e^- \rightarrow Fe + 2OH^-$	-0.19
$2H_2O + 2e^- \rightarrow H_2 + 2OH^-$	+0.037
$Cd(OH)_2 + 2e^- \rightarrow Cd + 2OH^-$	-0.143
$FeCO_3 + 2e^- \rightarrow Fe + CO_3^{2-}$	-0.422
$Co(OH)_2 + 2e^- \rightarrow Co + 2OH^-$	-0.193
$2SO_3 + 3H_2O + 4e^- \rightarrow S_2O_3^{2-} + 6OH-$	-0.27
$Fe(OH)_3 + e^- \rightarrow Fe(OH)_2 + OH^-$	-0.27

$$PbCO_3 + 2e^- \rightarrow Pb + CO_3^{2-} \qquad\qquad -0.423$$
$$S + 2e^- \rightarrow S^{2-} \qquad\qquad -0.06$$
$$S_4O_6 + 2e^- \rightarrow 2S_2O_3 \qquad\qquad -0.24$$
$$Pt(OH)_2 + 2e^- \rightarrow Pt + 2OH^- \qquad\qquad -0.273$$
$$O_2 + 2H_2O + 4e^- \rightarrow 4OH^- \qquad\qquad -0.809$$
$$MnO^{4-} + 2H_2O + 3e^- \rightarrow MnO_2 + 4OH^- \qquad\qquad -0.907$$

* Data were taken from *Electroanalytical Chemistry Basic Principles and Applications* by James A. Plambeck, Wiley, New York, 1982

Example 2.4. Determine the E^o value at 55 °C for the following electrode reaction: $I_2 + 2e^- \rightarrow 2I^-$

Solution We see from Table 2.2 that E^o value for this reaction is 0.5355 V which is of course a value at 25 °C and from Table 2.4 we obtain the temperature coefficient of -0.148 mV/K for the reaction. So, using equation 2.16 we get

$$E^o_{55} = 0.5355 \text{ V} - 0.148 \times 10^{-3} \text{ V/K}(382.16-298.16)K$$
$$= 0.5231 \text{ V}.$$

2.10 Heat of Reaction and Cell Potential

The Nernst equation explicitly implies that electrochemical cell potential is temperature dependent. The temperature coefficient of some cell potentials are listed on Table 2.4. It can be shown, from elementary thermodynamics, that the heat of reaction in an electrochemical reaction can also be obtained. Consider the Gibbs-Helmholtz equation:

$$\Delta G = \Delta H - T\Delta S \qquad\qquad 2.17$$

ΔH is the enthalpy or heat of reaction and ΔS is entropy (degree of randomness or disorder). At constant pressure, which is typical in most electrochemical reactions, the change in entropy is known to be a function of the change in Gibbs free energy change per change in temperature as shown in equation 2.18

$$\Delta S = -d\Delta G/dT \qquad\qquad 2.18$$

Since it has been shown that $\Delta G = -nFE$, we can then write the ΔS in terms of E thus:

$$\Delta S = -d(nFE)/dT = nF(dE/dT) \qquad 2.19$$

If we substitute equation 2.19 into equation 2.17 we obtain

$$-nFE = \Delta H - nFT(dE/dT) \qquad 2.20$$

Dividing equation 2.20 by $-nF$ will give

$$E = \Delta H/nF + T(dE/dT) \qquad 2.21$$

This equation is very useful in the sense that it can be used to estimate the electrochemical heat of reaction at a given temperature if the temperature coefficient of the reacting system is known.

2.11 Electrochemical Overpotential

Earlier in this chapter it was thermodynamically shown that for a spontaneous reaction, the E_{cell} of a given electrochemical reaction is positive and for electrolytic reaction it is negative. Suppose in an electrochemical reaction the theoretically calculated E_{cell} is negative and we still insist on making the reaction go. Thermodynamics tells us that potential in excess of the theoretical equilibrium potential (voltage) must be impressed across the cell for such a reaction to proceed. This excess potential, necessary to initiate the reaction, is referred to as overpotential, with the symbol, η. Therefore, η can be mathematically stated as follows:

$$\eta = E - E_{eq}$$

The consequence of this relation is that E is always less than E_{eq} and η is therefore always negative.

We have just used the word theoretical potential. Does this mean that the calculated cell potential can be different from the observed cell potential? The answer is "yes". In our consideration of cell potentials we ignored the fact that the solution through which ions migrate have a certain measurable resistance which, when multiplied with the associated current, gives voltage. This voltage is referred to as IR drop, which must be subtracted from theoretical potential to obtain the actual potential of the cell.

Therefore, the observed cell potential may be calculated as follows:

$$E_{cell} = E_{cathode} - E_{anode} - IR \qquad\qquad 2.22$$

The consequence of equation 2.22 is an increase in the potential necessary to operate an electrolytic cell or a decrease in the voltage output of a galvanic cell.

Other factors, which affect a theoretical cell potential, include liquid junction potential, which can be minimized by use of appropriate salt bridge and concentration polarization. Concentration polarization arises when there exists a slow rate of materials (electroactive materials) transport to the electrode to maintain the current as demanded by the reacting system.

2.12 Current Efficiency

In an electrolytic reaction we desire that the amount of charge passed be equal to the quantity of material produced or consumed. This is in line with Faraday's Law. If in actual fact this happens, we conclude that the reaction proceeded with 100% current efficiency.

More often than not this does not occur. Several factors such as the evolution of gas during electrodeposition can decrease the current efficiency of a given metal deposition. An example will illustrate the concept of current efficiency as it applies to electrochemical systems.

Example 2.5

At STP electrolysis of water was conducted by passing a current of 5.0 amps for 1 hour. A total of 960 mL of oxygen was collected. Calculate the current efficiency of the system.

Solution

Consider the following reaction for water decomposition:

$$2H_2O \rightarrow 4H^+ + O_2 + 4e^-$$

This equation states that for every mole of oxygen produced, 4F are involved in the reaction. But at STP one mole of oxygen gas occupies a volume of 22400 mL.

$$\therefore 960 \text{ mL}/22400 \text{ mL/mole} = 0.043 \text{ mole } O_2$$

1 mole of O_2 consumes 4F x 96500 Coul/F
0.043 mole O_2 consumes 4F x 96500 Coul/F x 0.043/1
$$= 16,598 \text{ Coul.}$$

This is the amount of charge transferred during the production of 960 mL of O_2. However, theoretically the expected charge is:

Q = it = 5.0 amp x 60 min x 60 sec/min = 18000 Coul.

A comparison of this theoretical charge with the actual charge of 16,598 Coul will show that the cell is not operating with 100 % current efficiency.

The current efficiency of this cell or reaction is:

$$(16,598 \text{ Coul}/18,000 \text{ Coul}) \text{ x } 100 = 92.21 \%.$$

2.13 Cell Voltage and Electrical Work

At the beginning of this monograph we defined electrochemistry as the study of the inter conversion of electrical and chemical energies. From the discussions thus far we have shown the relationship between the Gibb's free energy change during a chemical reaction and an electrochemical cell potential. We have also seen that a measurable amount of charge is produced when current is passed through a solution via conductors (electrodes). From thermodynamic consideration, electrical work, $w_{electric}$, is related to electrochemical type of work as shown in equation 2.23.

$$w_{electric} = -Q\Delta E = w_{max} \qquad \qquad 2.23$$
$$= -it\Delta E$$

The negative sign in this equation arises from thermodynamics in the sense that work done by a system on its environment is usually assigned a negative value from the $w = -p_{ext}dV$ in the p-V work. ΔE for galvanic cells is positive while it is negative for electrolytic cells. $w_{electric}$ is also the maximum electrical work performed when a galvanic cell is operating reversibly.

57

CHAPTER 3

3.0 SELECTED ELECTROCHEMICAL APPLICATIONS

So far we have considered some fundamental theories governing electrochemistry and electrochemical processes. In this chapter we shall look at some applications of these processes. Since this monograph is intended as an elementary introduction to the field, it is certainly beyond its scope to treat this chapter from a mathematical and mechanical viewpoints. Although approaches from these viewpoints are very important, it is believed that the sacrifice of a mathematical rigor will not impair the understanding of the materials presented here.

3.1 Storage Cells

Storage cells are electrochemical cells in which an external electric current is used to charge up the chemical materials in the cells. These cells operate galvanically. There are two kinds of storage cells: the primary and the secondary cells.

3.1.1   The Primary Cells

The primary cells are those cells that are used only once. They are not recharged. The container of this type of cell is part of the electrode material. An example of the primary cells is the well-known flash light battery. The most important cell of this type is the Lechlanche' or Mag cell. It is made up of a carbon rod covered with a paste of manganese dioxide, ammonium chloride and graphite. Amalgamated zinc is used as cover. The components of this cell are shown in Fig. 3.1(a) and the assembled cell is shown in Fig. 3.1(b).

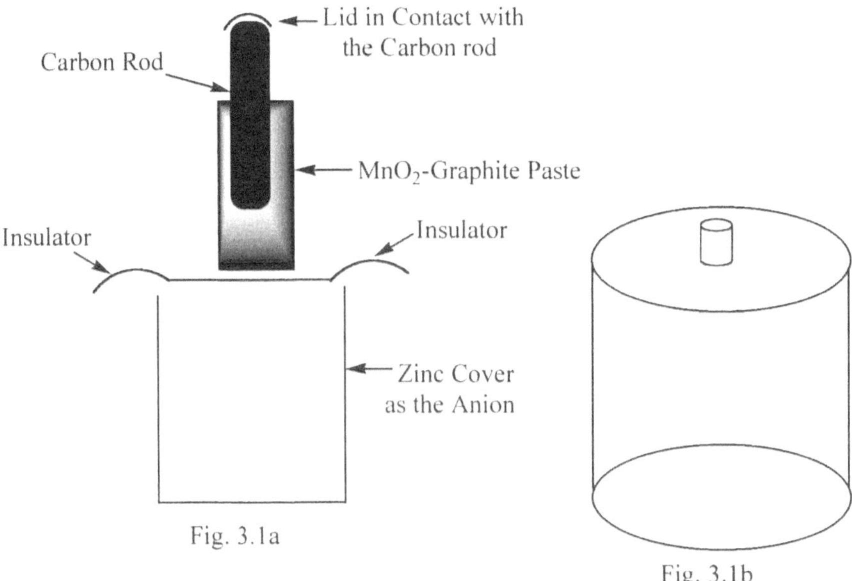

Fig. 3.1 Schematics of the Primary Cell Components and Assembly

The cell reaction of the Lechlanche′ cell is as follows:

$$\text{Anode: } Zn + 2OH^- \rightarrow Zn(OH)_2 + 2e^- \qquad E^o = +1.245 \text{ V}$$
$$\text{Cathode: } 2MnO_2 + H_2O + 2e^- \rightarrow Mn_2O_3 + 2OH^-$$
$$E^o = +0.27 \text{ V}$$

Overall cell reaction is:

$$Zn + 2MnO_2 + H_2O \rightarrow Zn(OH)_2 + Mn_2O_3$$

The E^o value of this cell is, 1.515 V. This means that, in operation, this cell is expected to deliver a potential of about 1.52 V. This type of cell is not rechargeable because during operation the Zn cover is oxidized to Zn^{2+} which complexes with other side reactions of the cell to produce insoluble crystalline deposits. These solid deposits render the cell passive and therefore produces no current. The formation of the crystalline products from the cell reaction byproducts may be represented as follows:

$$OH^- + NH_4^+ \rightarrow H_2O + NH_3$$
$$2NH_3 + Zn^{2+} + 2Cl^- \rightarrow Zn(NH_3)_2Cl_2 \text{ (solid)}$$
$$Zn^{2+} + 2OH^- \rightarrow H_2O + ZnO \text{ (solid)}$$

3.1.2 Secondary Cells and Batteries

Secondary cells are those storage cells that can be recycled many times. In other words, secondary cells are rechargeable. An example of a secondary cells is the lead-acid battery. However, before a discussion on batteries it will be appropriate to define what a battery is.

A battery may be defined as a collection of cells joined together by a conducting wire or metal strip to produce a cumulative voltage. Therefore, a cell can be regarded as a building block of a battery. If it is experimentally determined that a given cell has a potential of 1.2 V, it is possible to obtain as much as 12.0 V of energy if about 10 such cells are assembled and connected using a conducting wire as described above.

Batteries are very important in every facet of our everyday life. Flash light cells, powering of automobiles, pace makers, fuel cells, household electronic gadgetry and ancilliaries and even power generation in most industries and power stations, all function on batteries. This is more so in the field of the myriad industrial electronic devices that are constantly and continuously developing.   There is pressure on battery scientists to develop batteries to keep up with new electronic devices.   This fact was made quite clear recently in an article BUILDING A BETTER BATTERY by Joe Alper which appeared in the *Chemistry* magazine (Autumn 2002) in which John Hadley of Rayovac was quoted as saying that

> *Device manufacturers have forced us to push the limits of today's battery technologies and accelerate the development of the next generation of technologies. There are personal electronic devices that have already been developed but that need better batteries from us before consumers will get to play with them. – John Hadley*

As stated in the above article, "A better battery is one that delivers more power for a longer time from a smaller, lighter package – perhaps even at lower cost ...". In addition to this statement, figures of merit for good batteries are energy density and power density.

$$H_2 + \tfrac{1}{2}O_2 \quad \rightarrow \quad H_2O \quad E^o = 1.23 \text{ V}$$

Energy density of a battery may be defined as the stored energy per unit weight or volume of the battery materials. On the other hand, the rate of supply of energy per unit mass of a battery is the power density, given mathematically as:

$$\text{Power Density} = I_{max} \times V/\text{mass}$$

Where I and V are current and voltage across a battery cell, respectively. The performance of a battery is judged by these figures of merit and is limited by the electrode materials and the kinetic considerations of the redox reactions in the battery.

There are in existence different types of batteries such as the lead-acid, nickel-cadmium, silver-zinc. All these are storage or accumulator batteries.

They operate on the same principle of oxidation-reduction of the anode and cathode materials, respectively. They can be recycled many times.

Lead-acid battery is probably the most commonly known accumulator. It consists of two porous lead electrodes whose pores are filled with lead sulfate. The electrolyte in the lead-acid battery is sulfuric acid with lead sulfate. It is instructive to note that when these electrodes are coupled, no work will be done. As the name implies, it is an accumulator in the form of chemical energy.

During the charging process, the cathodic electrode reduces the lead ion (Pb^{2+}) which then deposits on the lead sheet (electrode) thereby making more lead. At the same time, the anodic electrode oxidizes lead ions which gives up electrons to the electrode and becomes Pb^{4+}. This lead (IV) can then undergo some hydrolysis reaction to form PbO_2. With this, the charging and discharging processes may be represented thus:

Charging Cycle
$$Pb^{2+} + 2e^- \rightarrow Pb \qquad \text{(negative electrode)}$$
$$Pb^{2+} + 2H_2O \rightarrow PbO_2 + 4H^+ + 2e^- \qquad \text{(positive electrode)}$$

This is actually the situation when a car battery is charged. That is, enough solid lead oxide (PbO_2) is deposited on one electrode, and on the other electrode, enough solid lead (Pb) is deposited. When a car is driven, the battery discharges, meaning the reverse reaction of the charging process occurs as shown by the reaction

$$\text{Discharging Cycle}$$

$Pb \rightarrow Pb^{2+} + 2e^{-}$ (positive electrode)

$PbO_2 + 4H^{+} + 2e^{-} \rightarrow Pb^{2+} \; 2H_2O$ (negative electrode)

It will be noted in the above equation that ions are used. In actual fact and for completeness of chemical processes occurring during cell reaction, the following is customarily used:

$$PbSO_4 + 2e^{-} \rightarrow Pb + SO_4^{2-} \qquad E^{o} = -0.28 \text{ V} \quad (1)$$

$$PbSO_4 + 2H_2O \rightarrow PbO_2 + 4H^{+} + SO_4^{2-} + 2e^{-}$$
$$E^{o} = +1.70 \text{ V} \quad (2)$$

Overall Cell Reaction

$$2PbSO_4 + 2H_2O \rightarrow Pb + PbO_2 + 4H^{+} + 2SO_4^{2-}$$

When the car is driven, the cell discharges spontaneously and as seen in equation (1), the Pb, which is known as the lead electrode, behaves as an anode while in equation (2) the lead oxide electrode is the cathode. The cell voltage can be evaluated as done earlier in the theoretical treatment of electrochemical cell reaction, that is,

$$E_{cell} = E_{cathode} - E_{anode}$$

Therefore, in operation each cell in lead-acid battery will deliver a potential of 1.70 V $-(-0.28$ V$) = 1.98$ V. For a regular car battery, six cells will therefore be required to produce the conventional 12.0 V. Figure 3.2 depicts a typical battery assembly.

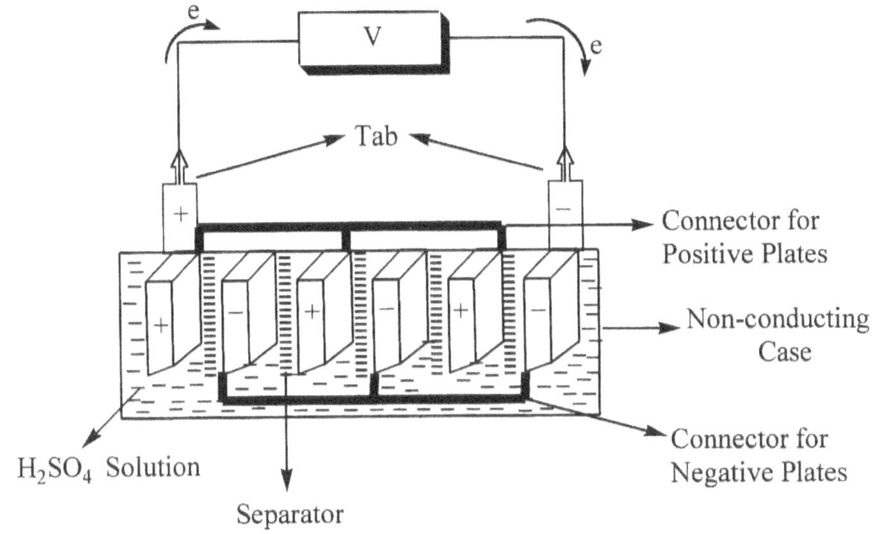

Fig. 3.2 Schematic Representation of Accumulator Battery

Battery chemistry is quite fascinating and its fabrication and ever increasing application make it worthwhile to give a listing of the different batteries in current use and their respective chemistries.

Table 3.1 Current Batteries in Use and their Chemistry

Primary Cells

Silver-zinc Cells
Anode: Zin powder
Cathode: Silver oxide slurry
Electrolyte: Aqueous KOH
Half reactions:

$$Zn + 2OH^- \rightarrow ZnO + H_2O + 2e^-$$
$$Ag_2O + H_2O + 2e^- \rightarrow 2Ag + 2OH^-$$

Alkaline Cells
Anode: Zinc powder
Cathode: Manganese dioxide powder
Electrolyte: KOH

Alkaline Cells
Half reactions:
$$Zn + 2OH^- \rightarrow ZnO + H_2O + 2e^-$$
$$2MnO_2 + H_2O + 2e^- \rightarrow Mn_2O_3 + 2OH^-$$

Zinc-Air Cells

Anode: Amalgamated zinc powder and electrolyte
Cathode: Oxygen (O_2)
Half reactions:
$$Zn + 2OH^- \rightarrow Zn(OH)_2 + 2e^-$$
$$1/2O_2 + H_2O + 2e^- \rightarrow 2OH^-$$

Secondary Cells

Lead-Acid Battery

Anode: Sponge metallic lead
Cathode: Lead dioxide (PbO_2)
Electrolyte: H_2SO_4
Half-Cell Reaction:
$$Pb + SO_4^{2-} \rightarrow PbSO_4 + 2e^-$$
$$PbO_2 + SO_4^{2-} + 4H^+ + 2e^- \rightarrow PbSO_4 + 2H_2O$$

Sodium-Sulfur Cells

Anode: Liquid sodium
Cathode: Liquid sulfur
Electrolyte: solid electrolyte (porous aluminum oxide)
Half reactions:
$$2Na \rightarrow 2Na^+ + 2e^-$$
$$S + 2e^- \rightarrow S^{2-}$$

Nickel-Cadmium Cells

Anode: Cadmium
Cathode: Nickel hydroxide
Electrolyte: Aqueous KOH
Half reactions:
$$Cd + 2OH^- \rightarrow Cd(OH)_2 + 2e^-$$
$$NiO_2 + 2H_2O + 2e^- \rightarrow Ni(OH)_2 + 2OH^-$$

Nickel-Metal Hydride (NIMH) Cells
Anode: Rare-earth or nickel alloys with many metals
Cathode: Nickel hydroxide
Electrolyte: KOH
Half reactions:
$$MH + OH^- \rightarrow M + H_2O + e^-$$
$$NiOOH + H_2O + e^- \rightarrow Ni(OH)_2 + OH^-$$

Solid Cathode Lithium Cells
Anode: Lithium
Cathode: A heat-treated MnO_2
Electrolyte: Propylene carbonate and 1,2-dimethoxyethane
Half reactions:
$$Li \rightarrow Li^+ e^-$$
$$Mn^{IV}O_2 + Li^+ + e^- \rightarrow Mn^{III}O_2(Li^+)$$

Lithium Ion cells
Anode: Carbon compound
Cathode: Lithium oxide
Electrolyte: $LiPF_6$
 Chemistry: Based on "intercalation", the reversible insertion of guest atoms like lithium into host solids like the battery electrode materials

3.2 Fuel Cells

We may start discussion in this section by first deciding what a fuel is. Consider a hydrocarbon, say, methane, CH_4. The following thermochemical equation shows the fate of this hydrocarbon as well as the accompanying energy released when it is burned in air.

$$CH_4 + 2O_2 \rightarrow CO_2 + 2H_2O \qquad \Delta H = -890.0 \text{ kJ}$$

What does this mean? It means that whenever we burn 1 mole or 16.0 g of natural gas in air, an energy output of 890.0 kJ is obtained. This energy can be used to perform some work. The substance that produced this energy may be looked at as a source of fuel. Therefore, a fuel is any substance that can be combusted to produce energy. Fuel cells, in the same way, may be defined as a system that can be used to combust or oxidize an energy-producing material.

The most common of the fuel cells is the hydrogen-oxygen fuel cell whose reaction may be represented as:

$$H_2 \ + \ 1/2O_2 \ \rightarrow \ H_2O \qquad\qquad \Delta H = -285.77 \text{ kJ}$$

However, this process does not occur spontaneously at ambient temperature. Therefore, in order to harness this much energy a catalyst must be used to ensure rapid reaction. The catalyst that is most often used in fuel cell technology is the porous nickel electrode.

We have so far referred to porous lead and nickel electrodes. How are they made? An example of the preparation of porous nickel electrode will suffice to educate a layman the principles and meaning of the term "porous" in electrochemical systems.

A nickel porous electrode is prepared by several methods, one of which is by nickel powder method. By this method, a nickel powder is pressed on a wire screen. When the screen is covered to a desired depth, the whole set-up is subjected to very high thermal energy (in a furnace) so that the loose nickel powder forms a continuous body with the screen. Other processes may follow but the end result is a porous plate. Electrochemical fuel cell is obtained by coupling two porous electrodes making one an oxygen electrode and the other, a hydrogen electrode. Fig. 3.3 illustrates a typical fuel cell set up.

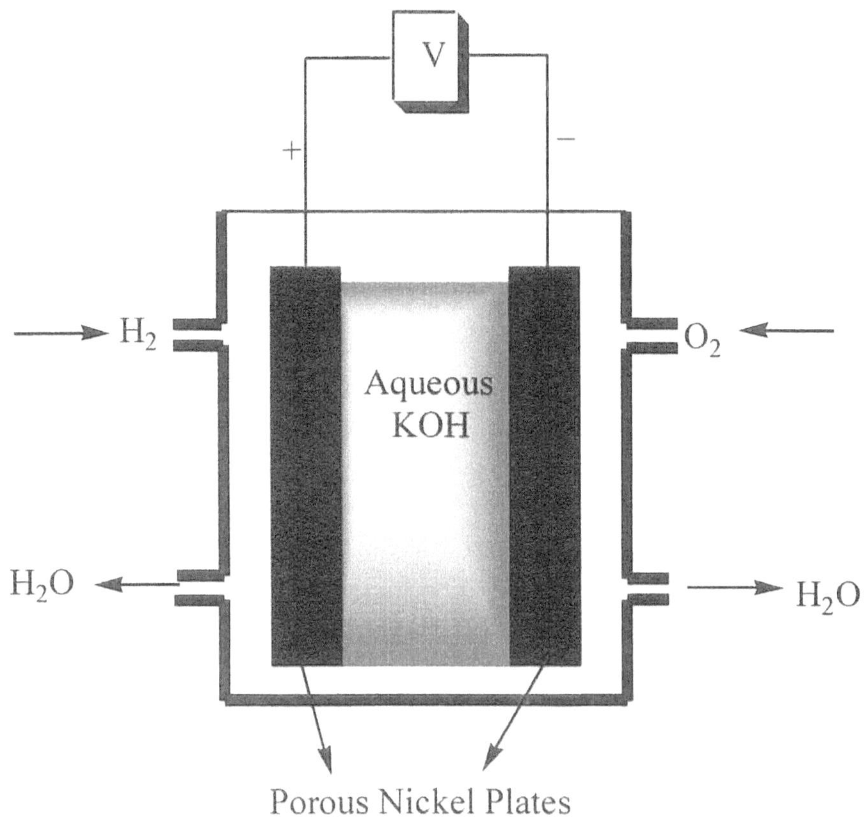

Fig. 3.3 Fuel Cell Set Up

The reaction taking place at these electrodes are:

$$\text{Anode:} \quad H_2 \rightarrow 2H^+ + 2e^- \qquad E^o = 0.00 \text{ V}$$
$$\text{Cathode:} \quad 1/2 O_2 + 2H^+ + 2e^- \rightarrow H_2O$$
$$E^o = 1.23 \text{ V}$$

The overall cell reaction is:

$$H_2 + \tfrac{1}{2}O_2 \rightarrow H_2O \quad E^o = 1.23 \text{ V}$$

The voltage produced by a single cell following the above reaction is therefore 1.23 V. The produced water is continuously removed and may be used to supplement drinking water, especially in space travels.

3.2.1 **Water** Electrolysis

This discussion will not be complete without mentioning the electrolysis of water which operates in the same line as fuel cell except that the feed-stocks in both reactions are different. While the feed-stock for fuel cell is oxygen and hydrogen, the feed-stock for water hydrolysis is water with an added electrolyte.

Decomposition of water to yield hydrogen and oxygen is a further example of electrolysis. This technology has been of tremendous interest nowadays because of the need for renewable and alternative energy source. Hydrogen has been found to be a useful source of energy that can be obtained by the electrolysis of pure water, as can be seen by the following equation:

$$2H_2O(l) + 2e^- \quad \rightarrow \quad 2H_2(g) \; + \; O_2(g) . \quad E^o = -1.229 \text{ V}$$

 Hydrogen thus produced can be appropriately stored for future use. Of course, water does not conduct very well. In other to use this technology a suitable electrolyte that is soluble in the water has to be used. The electrolyte that is commonly used is H_2SO_4 which dissociates in water thus:

$$H_2SO_4 \quad \rightarrow \quad 2H^+ + SO_4^{2-} + 2e^-$$

When in water, the SO_4^{2-} is oxidizes at the anode as:

$$2SO_4^{2-} \quad \rightarrow \quad S_2O_6^{2-} + 2e^- \quad E^o = -0.22 \text{ V}$$

Anion with such low E^o will not be in competition with the anion produced by the dissociation of water, OH^-, whose E^o is -0.83 V. The arrangement for the electrolysis of water is shown below which is akin to the Hoffman voltmeter used in the electrolysis of water.

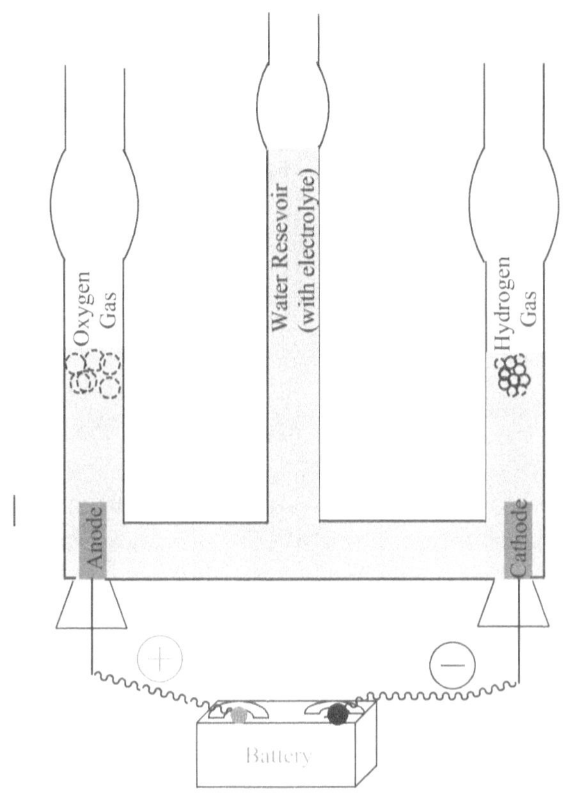

Fig. 3.3.1 The Set Up of Hoffman Voltmeter for Water Electrolysis

3.3 Corrosion

Corrosion may be defined as a spontaneous destruction of metals by the environment with which it is in contact. Such environments may be atmospheric, chemical, electrochemical or biochemical, and this undesirable destruction causes quite a substantial financial loss each year. It is estimated in 1998 that the annual cost of corrosion in the United States of America was about \$121.41 billion, which is about 1.381 % of the Gross Domestic Product (GDP).   To understand the process of corrosion, we may start with a given metal, say, iron, whose rusting process may be rightly called corrosion. Metals are usually obtained in a combined state, and in order to convert them into a free state, some energy

69

is required. We may do a simple analysis of converting iron that is obtained in nature as Fe_2O_3 into its free form, Fe. The process of this conversion involves several steps, but from an electrochemical standpoint, the following equations may suffice for the point we intend to illustrate. Fe_2O_3 is in +3 oxidation state and may be written in the following form:

$$2Fe^{3+} + 6e^- \rightarrow 2Fe \qquad\qquad E^o = -0.036\ V$$
$$3H_2O \rightarrow 6H^+ + 3/2O_2 \qquad\qquad E^o = -1.229\ V$$

Overall Cell Reaction:

$$2Fe^{3+} + 3H_2O \rightarrow 2Fe + 6H^+ + 3/2O_2 \quad E^o = -1.265\ V$$

The E^o values used in these reactions imply, as usual, that the process is occurring at standard temperature of 25°C. Now we ask if this process can occur spontaneously. The answer is NO – because the ΔG of the process is positive, that is

$$\Delta G^o = -nFE^o = -6 \times 96500\ C \times (-1.265\ V) = 73.244\ kJ.$$

The consequence of this is that when iron is in contact with its environment containing oxygen and moisture, a spontaneous redox reaction occurs, whereby the iron, which has a higher tendency to revert to its oxidized form, gives up electrons which are accepted by oxygen. When this process occurs even a cursory observation, will reveal the change in appearance of the iron to be brownish and we will say that it is corroded and others may say that it is hydrolyzed. The brownish compound is iron in its +3 oxidation state. This process is true and typical of the process of corrosion of metals that may, in general, be represented as:

$$M + H_2O + 1/2O_2 \rightarrow M(OH)_2$$
$$OR$$
$$M + 3/2H_2O + 3/4O_2 \rightarrow M(OH)_3$$

etc., depending on the oxidation state of the metal, M.

Let us consider the corrosion of a metal, say, Zn to Zn^{2+}. The corrosion of Zn to Zn^{2+} is depicted in Fig. 3.4.

Zn is losing electrons to become Zn^{2+} and the electrons are accepted by O_2 which reacts with H^+ to produce H_2O. Then the Zn^{2+} ions produced react with water to produce $Zn(OH)_2$ which deposits. In time, the active area eats away and the sheet is destroyed.

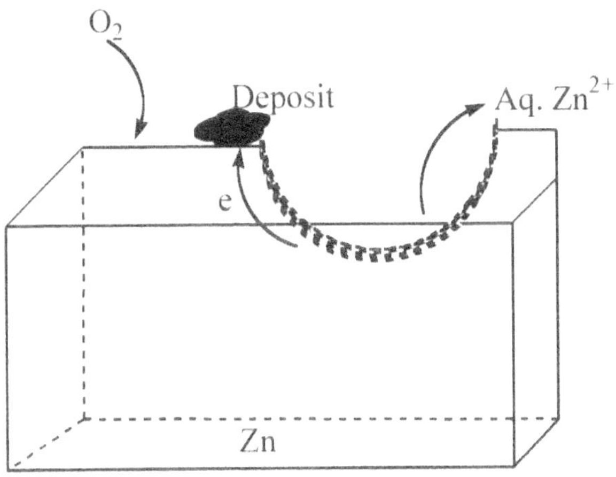

Fig. 3.4 Schematic of Zn Corrosion

It will be appreciated that the principle of "differential aeration" is in operation in this example. That means that the area where there is oxygen is preferentially reduced and the area where there is oxygen starvation acts as the anode. The means that part of the metal exposed to the oxygen-starved environment will be the part where the destructive oxidation (corrosion) takes place.

The example given above is not the only way that metals can corrode. Even a metal in contact with another metal will corrode. The E^o value of the metals will determine which of the metals in contact will corrode. For example, suppose we have iron, $E^o = 0.4402$ V, in contact with zinc, $E^o = -0.7628$ V and we wish to determine which of these metals will corrode. We note that corrosion is a spontaneous process and as such, the free energy change must be negative. For this to be the case, the E^o_{cell} of the two metals must be positive, that is,

$$E^o_{cell} = E^o_{cathode} - E^o_{anode}.$$

Therefore, E^o for these two metals must be:

$$-0.4402 \text{ V} - (-0.7628 \text{ V}) = +0.3226 \text{ V}.$$

This results means that the anode metal must be zinc while iron will be the cathode metal. As a result, we may write the reaction thus:

$$Zn \rightarrow Zn^{2+} + 2e^-$$
$$\underline{Fe^{2+} + 2e^- \rightarrow Fe}$$

$$\text{Reaction: } Zn + Fe^{2+} \rightarrow Zn^{2+} + Fe$$

We now check on the free energy change in this process.

$$\Delta G = -nFE = -2 \times 96500 \text{ C} \times 0.3226 \text{ V} = -6.23 \text{ kJ}$$

The negative free energy assures spontaneous corrosion of zinc when in contact with iron. The principle, demonstrated by this example, is used in the selection of "sacrificial anode" material in most corrosion prevention practices. A discussion on "sacrificial anode" is presented later.

3.3.1 Corrosion Control

Corrosion of metals is very costly and most undesirable, and it occurs everyday. Consequently, corrosion scientists have tried several methods of preventing and, in some extreme cases, minimizing the rate of metal corrosion. Although the methods used are many and only a few will be discussed here, they can be subdivided into two main categories: Protective Coating and Corrosion Inhibitors.

Protective Coating is the coating of a substrate material (metal) with organic compounds such as polymers (polyurethane, polyvinyl chloride (PVC), Nylon, polytetrafluoroethylene (PTFE)), some long chain quaternary ammonium salts, amides and metals (for example, alloying and galvanization. In this method a wet coating of the inhibitor is applied on the substrate and allowed to cure (usually by evaporation of the coating solvent) leaving a protective coat of the inhibitor on the metal substrate. On the other hand, Corrosion Inhibitor is a substance that reduces the rate of metal oxidation in a given environment. Often the concentration of the inhibitor is low. Therefore, the mechanism in this method of corrosion inhibition is the formation of a protective passive film of the inhibitor on the surface of the metal substrate.

3.3.2 Painting

Painting, which is a form of coating, may be used to protect a metal. The paint prevents O_2 and moisture and other deleterious materials from getting to the metal of interest. This way corrosion is prevented. However, if the paint chips away from the metal surface, corrosion of the metal commences.

3.3.3 Galvanization

This is a form of painting. Coating iron with a thin layer of zinc produces galvanized iron. The zinc protects the ion from corrosion. Even after the zinc is scratched and a surface of iron is exposed, the zinc, which has a higher tendency to oxidize than iron, acts as the anode and is preferentially oxidized (corroded). In the protection of iron buried underground, a connection is made between it and a zinc rod or pole. The zinc now acts as the anode and is also preferentially oxidized. It is therefore referred to as a 'sacrificial' anode. Any other metal of higher E^o value than iron may also be used as a sacrificial anode for iron protection.

3.4 Electrodeposition

Electrodeposition is a general term that may be applied to electroplating, electroseparation, electrowinning, etc. The general principle involved in electrodeposition is the selection of a potential at which the metal of interest can be deposited at a cathodic electrode. This electrochemical method can also be used in quantitative chemical analysis where it is referred to as electrogravimetry.

Suppose we wish to refine a metal say Cu from solid CuS. All that is required of us to do is to use the solid CuS as our anode material and then obtain a pure Cu sheet and make it a cathode. These two electrodes are now immersed in a solution of $H_2SO_4/CuSO_4$. From our knowledge of electrochemical reaction so far, we know that as the current passes through the cell, the anode material dissolves producing Cu^{2+} and S^{2-}. But the question is, what is plated on the cathode, Cu or S? The answer lies in the following reactions:

$$Cu^{2+} + 2e^- \rightarrow Cu \qquad E^\circ = + 0.337 \text{ V}$$
$$S^{2-} \rightarrow S + 2e^- \qquad E^\circ = + 0.480 \text{ V}$$

$$\text{Overall: } Cu^{2+} + S^{2-} \rightarrow Cu + S \qquad E^\circ = + 0.817 \text{ V}$$

The reaction as written proceeds spontaneously and Cu^{2+} is seen to be more easily reduced than S^{2-}. Therefore, the dissolved Cu^{2+} will plate on Cu cathode. Copper produced by this method is usually 99.999 % pure. Had the process been non-spontaneous, then a potential in excess of the calculated value would have to be imposed on the cell to effect the desired result.

3.5 Potentiometry and Ion Selective Electrode

The use of electrical signal for quantitative determination of a given analyte in a solution is a very important segment of the broad area of analytical chemistry. Both current and potential generated at an indicator electrode are useable and have been used as sensitive analytical signals. Potentiometry is the use of electrical potential, produced by ions in solution, to determine the activity or concentration of such ions. The potential so observed is usually proportional to the activity or concentration in the solution following the Nernst equation.

$$E = E^\circ - 0.059/n \log a_i$$

The importance of this technique is that when an electrode, sensitive to the activity of a given ion, is immersed into a solution of such ion, a potential develops. This potential could then be used to determine the concentration or activity of that ion.

Electrode materials can be fabricated to respond to specific ions in solution following the potentiometric method. Such electrodes are referred to as ion selective electrode (ISE). Many such electrodes have been fabricated and are used in almost all our daily activities, ranging from simple pH measurement, using a pH electrode (electrode sensitive to the hydrogen ion) to calcium ISE for direct measurement of calcium ions in biological fluids, in soils and in water.

Currently, about twenty ISEs have been developed to measure the activity (concentration) of specific ions or even neutral molecules such as CO_2 and NH_3. The Beckman Handbook of Applied Electrochemistry gives a comprehensive list of ISEs in current use, their application and useful concentration range.

3.6 Electrosynthesis

Electrosynthesis is becoming a very versatile preparative method for chemical synthesis. The principle involved in electrosynthesis is a careful selection of potential at which the reacting material is either reduced or oxidized, depending on which compound is desired. One of two methods for obtaining this potential is by literature survey to see if the E^o of such compound is known. If it is known, usually an overpotential of about 0.20 V to 0.50 V is added to the literature value. The reason for over potential has already been explained.

If, on the other hand, the E^o value of the compound of interest is unknown or it is not readily available, then a rapid determination is made to obtain the same by an i-E measurements. An example of this method will make this a little clearer. Suppose that it is desired to carry out an electrolytic synthesis of a compound, Y, in acidic medium. The first step is, of course, an electrochemical set up shown in Fig. 3.5.

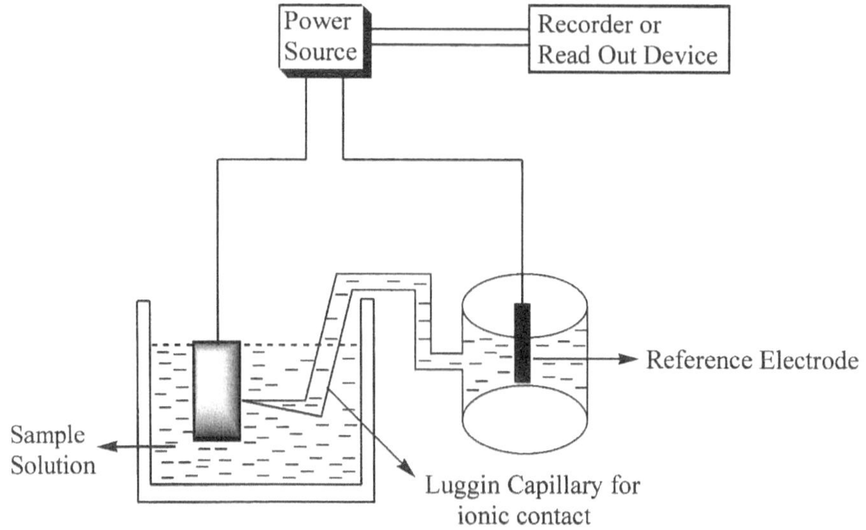

Fig. 3.5 Preparative Electrolysis Cell Set Up

When a suitable amount of this sample is dissolved in the reaction vessel and a variable potential is applied to the solution, an i-E curve is usually obtained. This is shown in Fig. 3.6.

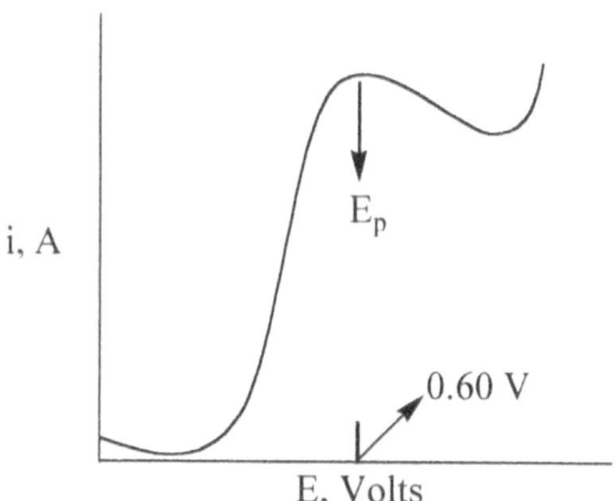

Fig. 3.6 Typical i-E Curve

At the point E_p, which is called peak potential, the compound is maximally oxidized, if oxidation is the process required.   Therefore, potential selection for electrosynthesis will be at the potential corresponding to E_p. If this compound is subjected to an applied potential, say 0.6 V, and the resulting current is monitored with respect to time, the following i-t curve is obtained:

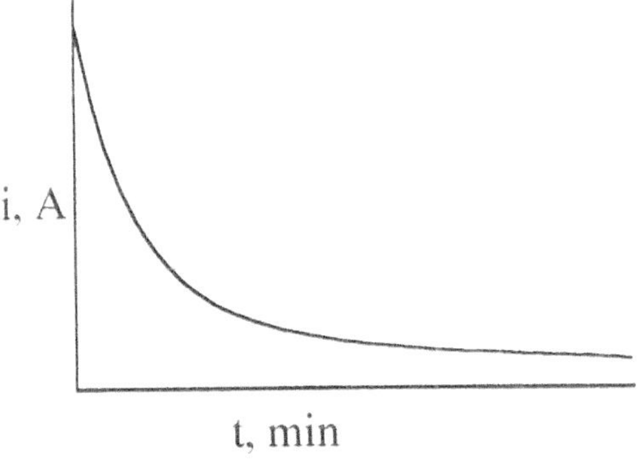

Fig. 3.7 i-t Curve at Constant Potential

As can be seen, this curve shows an exponential decay of the current with time, which can be quantified with the equation:

$$i = i_o e^{-kt}$$

where k is a characteristic decay constant for that compound. Evaluation of this equation yields Q, (it), which is charge. As may be recalled from chapter 2, this is the charge passed during the oxidation or reduction of the compound of interest. Recall that

$$Q = it = nFN \qquad\qquad 3.1$$

where N is the number of moles of the oxidized or reduced compound. A knowledge of n, and a corresponding determination of the pH dependency of the reaction, will not only provide a quantitative information on the oxidized or reduced compound, it will also provide information on the mechanism of the reaction.

The following compounds that are of industrial importance have been synthesized electrochemically:

Electrosynthesis of:

a) Adiponitrile

$$2CH_2=CHCN + 2H^+ + 2e^- \rightarrow NCCH_2CH_2CH_2CH_2CN$$
$$\text{Adiponitrile}$$

Adiponitrile is used in the production of nylon.

b) Melamine

$$HCN + Br^- \rightarrow BrCN + 1/2H_2 + e^-$$
$$BrCN + 2NH_3 \rightarrow CNNH_2 + NH_4Br$$
$$3CNNH_2 \rightarrow C_3N_6H$$
$$\text{Melamine}$$

Melamine condenses with formaldehyde (CH_2O) to give a thermosetting resin, which is used in the manufacture of plastic dishes and coating.

The synthesis set-up and mechanisms involved in these electrosynthesis reactions are, in some cases, complex and certainly beyond the scope of this monograph. Their introduction here merely serves as an example of the utility of electrochemistry in organic synthesis.

3.7 Electrochemistry in Biological Systems

A branch of science that uses electrochemistry and electrochemical techniques to study biological systems has evolved over the past few decades. This branch of science is rightly called bioelectrochemistry.
Biological processes involve transfer of materials and charges across an interphase. Most of the interphases associated with biological systems are membranes. Electron transport/phosphorylation, chemiosmotic processes, photosynthesis and almost all membrane activities in biological systems involve membrane, and whenever there is a membrane or interphase, there must, almost invariably, exist a potential difference. This potential difference causes a driving force for biological activities and reactions and these activities will, of necessity, require a knowledge of electrochemistry to understand the complex energetics involved in these activities. This fact is succinctly presented in the following abstract from Tien

> *...electron transfer, redox reaction, ion and proton gradient, membrane potential and ATP synthesis. Obviously, these membrane-based activities, many of which are electrical in nature, should be probed and investigated by electrochemical methods ...*
>
> *H. Ti Tien.*

We start the study of this exciting, albeit, complex system, by stating the fundamental law governing the driving force of movement of materials and charges across membranes and interphases. A concentration gradient causes a movement of systems from a high to a low concentrated region, and the driving force responsible for this is referred to as chemical potential. In its simplest form, this potential is given in terms of free energy change stated in equation 3.2.

$$\Delta G = RT \ln C_2/C_1 \qquad\qquad 3.2$$

C_1 and C_2 in this equation are the concentrations of solutes in phases 1 and 2 or the concentrations of solutes of interest on the outside and inside of a membrane. However, in biological systems, both charged solutes and/or electrons are always involved, therefore the potential difference involved in the transport of these particles must be included as part of a driving force. As we earlier noted, a potential difference, ΔE, between interphases or membrane is characterized by the free energy change: $\Delta G = nF\Delta E$. When this quantity is added to the chemical potential equation, we obtain equation 3.3

$$\Delta G = RT \ln C_2/C_1 + nF\Delta E \qquad\qquad 3.3$$

Equation 3.3 is referred to as electrochemical potential, which accounts for the driving force of charged solutes to move across biological membranes or associated interphases.

We consider here a proton movement from the inner membrane to the outer membrane of a biological system (taken from the chemiosmotic theory of Peter Mitchell).

$$H^+_{in} \rightarrow H^+_{out}$$

The driving force for this movement is expressed using equation 3.3, that is,

$$\Delta G = RT \ln [H^+]_{out}/[H^+]_{in} + nF\Delta E$$
$$= 2.303RT \log [H^+]_{out}/[H^+]_{in} + nF\Delta E.$$

We note here that $-\log[H^+] = pH$ and for $n = 1$ the above equation is rewritten as

$$\Delta G = -2.303RT(pH_{out} - pH_{in}) + F\Delta E$$
$$= 2.303RT\Delta pH + F\Delta E$$

Typical values of ΔpH is -1 and $\Delta E = 0.18$ V. Taking a temperature of 37 °C = 310 K and inserting these values into the above expression we obtain

$$\Delta G = 2.303RT + 96500 \text{ C} \times 0.18 \text{ V} = 23.31 \text{ kJ}$$

ΔpH of -1 implies that the pH outside is lower than the pH inside a membrane as experimentally determined. As we have shown above, this positive ΔG means an unfavorable situation but the reverse process is favorable and spontaneous.

In chemical systems the formal redox potential of an electrode is referred to as E^o, with stipulated conditions under which they were measured. In biological systems the formal redox potentials are measured at 25 °C and at a neutral pH of 7.0. For this reason, the formal redox potentials are referred to as $E^{o'}$. The $E^{o'}$ of most biological compounds are known and few of those are listed on Table 3.2. It may be worthwhile to compare these values with those listed for chemical systems on Table 2.2.

Table 3.2 $E^{o'}$ Values of Selected Biological Systems‡

Reaction	$E^{o'}$, Volts
$\frac{1}{2}O_2 + 2H^+ + 2e^- \rightarrow H_2O$	0.816
$Fe^{3+} + e^- \rightarrow Fe^{2+}$	0.771
Photosystem P700	0.430
$NO_3^- + 2H^+ + 2e^- \rightarrow NO_2^- + H_2O$	0.421
Cytochrome $f(Fe^{3+}) + e^- \rightarrow$ Cytochrome $f(Fe^{2+})$	0.365
Cytochrome $a_3(Fe^{3+}) + e^- \rightarrow$ Cytochrome $a_3(Fe^{2+})$	0.350
Cytochrome $a(Fe^{3+}) + e^- \rightarrow$ Cytochrome $a(Fe^{2+})$	0.290
Rieske Fe-S$(Fe^{3+}) + e^- \rightarrow$ Rieske Fe-S(Fe^{2+})	0.280
Cytochrome $c(Fe^{3+}) + e^- \rightarrow$ Cytochrome $c(Fe^{2+})$	0.254
Cytochrome $c_1(Fe^{3+}) + e^- \rightarrow$ Cytochrome $c_1(Fe^{2+})$	0.220
$UQH^\cdot + H^+ + e^- \rightarrow UQH_2$ (UQ = coenzyme Q)	0.190
$UQ + 2H^+ + 2e^- \rightarrow UQH_2$	0.060
Cytochrome $b_H(Fe^{3+}) + e^- \rightarrow$ Cytochrome $b_H(Fe^{2+})$	0.050
Fumarate $+ 2H^+ + 2e^- \rightarrow$ Succinate	0.031
$UQ + H^+ + e^- \rightarrow UQH^\cdot$	0.030
Cytochrome $b_5(Fe^{3+}) + e^- \rightarrow$ Cytochrome $b_5(Fe^{2+})$	0.020
$[FAD] + 2H^+ + 2e^- \rightarrow [FADH_2]$	0.003-0.091*
Cytochrome $b_L(Fe^{3+}) + e^- \rightarrow$ Cytochrome $b_L(Fe^{2+})$	-0.100
Oxaloacetate $+ 2H^+ + 2e^- \rightarrow$ Malate	-0.166
Pyruvate $+ 2H^+ + 2e^- \rightarrow$ Lactate	-0.185
Acetaldehyde $+ 2H^+ + 2e^- \rightarrow$ Ethanol	-0.197
$FMN + 2H^+ + 2e^- \rightarrow FMNH_2$	-0.219
$FAD + 2H^+ + 2e^- \rightarrow FADH_2$	-0.219
Gluthathione (ox) $+ 2H^+ + 2e^- \rightarrow$ 2 Gluthathione (red)	-0.230
Lipoic acid $+ 2H^+ + 2e^- \rightarrow$ Dihydrolipoic acid	-0.290
$NAD^+ + 2H^+ + 2e^- \rightarrow NADH + H^+$	-0.320
$NADP^+ + 2H^+ + 2e^- \rightarrow NADPH + H^+$	-0.320
α-Ketoglutarate $+ CO_2 + 2H^+ + 2e^- \rightarrow$ Isocitrate	-0.380
$2H^+ + 2e^- \rightarrow H_2$	-0.421
Ferredoxin (spinach) $(Fe^{3+}) + e^- \rightarrow$ Ferredoxin (Fe^{2+})	-0.430
Succinate $+ CO_2 + 2H^+ + 2e^- \rightarrow \alpha$-Ketoglutarate $+ H_2O$	-0.670

‡Values taken from *Biochemistry* 2nd ed by Reginald H. Garrett and Charles M. Grisham, Saunders College Publishing, NY, 1999

* Typical values for reduction of bound FAD in flavoproteins

Ingenious electrodes and electrochemical techniques have been devised over the years to monitor and to provide enabling data for use in predicting biological processes and their accompanying energetics. In addition, these advances are helping to better understand biological events *in vivo* and *in vitro* and thereby aiding, in no small measures, advances in modern medicine and biotechnology.

CONCLUDING REMARK

This monograph is written with a view to enlighten the layman on the subject of the general principles of electrochemistry and its far reaching applications. Certain topics have been purposely omitted but such omissions do not in any way impair the intended objective.

I hope that this eye-of-the-bird description of the ever-increasing applications of this branch of science will arouse the interest of many people. This should be more so since electrochemical science has assumed a status of its own in its own right. In recognition of this, Departments of Electrochemistry and Electrochemical Engineering have been established in many universities.

It is hoped that the results and discoveries coming out of these research institutes and departments will usher in a greater and more rewarding future for the benefit of mankind.

MOI

85

PROBLEMS

1. Consider a strong electrolyte, KOH, of 1×10^{-2} M concentration. This electrolyte dissociates into K^+ and OH^- ions. What is the $[OH^-]$ in this solution?

   Answer 1×10^{-2} M.

2. A 0.01M of a weak acid HX whose Ka is 1x10-5 is dissolved in solution and the resulting pH is measured to be 5.7. Calculate the degree of ionization and the percentage ionization.

   Answer 5.0×10^{-4}, 0.002%

3. Calculate the ionic strength of a solution made up of 0.1 M Na2SO4 and 0.01 M K3PO4. What is the activity coefficient of SO42- in this solution? Assume that the effective ionic radius of SO42- is 4.0 Å.

   Answer 0.36, 0.68

4. Balance the following reaction using the half reaction method:
   $Cr_2O_7^{2-}(aq) + I^-(aq) \rightarrow Cr^{3+}(aq) + I_2(s)$. In this reaction the cell potential observed when $[Cr_2O_7^{2-}] = 0.5$ and $[I^-] = 0.13$ M and pH of the solution is 0 is 0.8 V, calculate the concentration of $Cr^{3+}(aq)$ in the cell. Use this information to estimate the equilibrium constant of this reaction.

   Answer
   $$Cr_2O_7^{2-}(aq) + 14H^+(aq) + 6I^-(aq) \rightarrow 2Cr^{3+} + 3I_2(s) + 21H_2O(l)$$
   8.65e-4 M, 0.31

5. If the conductance of the ions in the above solution is known to be 0.25 Ω-1, calculate the resultant voltage that will be obtained if a current of 2 amp is passed through it.

   Answer 10.0 V

6. A 5.0V is impressed across a given electrolytic solution with a resistance on 10.0 ohms. If the operation took 20 mins, calculate the amount of current that resulted.

   Answer 0.5 Amps.

7. Assume that H_2O can be completely electrolyzed to produce H_2 and O_2 gases. Calculate the volume of water that must be electrolyzed to produce 10.0L of O_2 at STP if 20.0 Amp of current is passed through the water for 2.0 hours. You may also assume that 1g of H_2O = 1mL.

Answer 43.09ml.

8. Calculate the molar conductivity of a 0.01-M solution of KCl at 25 °C if its specific conductance is 1.408×10^{-3} $\Omega\text{-1cm}^{-1}$.

Answer 0.1408 S-m^2/mol

9. Using Table 1.3 calculate the diffusion coefficient of aqueous K^+ ion. Estimate the transference number of this ion in solution of KCl.

Answer 1.96×10^{-5} cm^2/s, 0.49

10. Consider the following cell notation:
Cu/0.1MCuSO4/Ni
$E^o_{Cu/Cn2+} = 0.337$, $E^o_{Ni/Ni2+} = -0.25$
(a) As written, which electrode will be plated with Cu?
(b) Can the reaction go spontaneously? If not, why?

Answer (a) Ni
(b) No, because the overall E^o is
negative.

11. If the Ni in problem 6 is to be plated, what voltage is necessary to initiate the reaction as written?

Answer At least 0.7V.

12. Consider problem 6 and 7, how many grams of Cu will be plated out if a current of 0.5 amp is passed through the cell for 1 hour.

Answer 0.593g.

13. Estimate the E^o value that can be obtained at 35 °C for the following reaction: $Mg(OH)_2 + 2e^- \rightarrow Mg + 2OH^-$. Use Tables 2.1 and 2.3 for this problem.

Answer -3.43 V

If in problem 8 only 0.490g of Cu was plated at the given time, calculate the current efficiency of the system.

87

Answer 82.63%

14. Consider the following:
    Sn / 0.05M Sn^{2+} // 0.2M Sn^{4+} / Sn
    $E^{o}_{Sn/Sn^{+2}} = +0.136V$; $E^{o}_{Sn^{+2}/Sn^{+4}} = -0.15V$
    (a) As written, calculate the resultant E of this cell.
    (b) What is the K_c of the reaction?
    (c) Will the cell react spontaneously as written
    (d) If you want the cell to give about 3.0 V, what will you do?

Answer (a) 0.372 V, (b) 1.25 x 10^{-3}, (c) Yes

(d) Assemble 10 of the cells

15. Consider the reaction: $O_2 + 4H^{+} + 4e^{-} \rightarrow 2H_2O$. $E^{o} = 1.229$ V.
    Using the temperature coefficient given on Table 2.4 calculate the
    heat of formation of one mole of water. Does this agree with the
    literature value of $\Delta H^{o}_{f.H2O}$.?

Answer 285.85 kJ/mole, Yes.

17. A system is known to deliver 5.0 mV in a reaction in which 2
    electrons were transferred. The reaction is performed at 30 °C.
    Calculate the K_c and ΔG for this system.

Answer 1.47, 970.53 J

18. A 9.0-V battery delivers a steady current of 1.5 A for 3.5 hours.
    Calculate the quantity of charge passed during this period and the
    maximum electrical work performed by this battery.

Answer 1.89 x 10^{4} C; -1.70 x 10^{5} J

19. A constant potential electrolysis was performed on a solution of
    MB for one hour during which period a total current of 500 mA
    was passed. If it is known the number of electrons involved in this
    reaction is 2, calculate the number of moles of MB that was
    electrolyzed.

Answer 9.33 x 10^{-3} mol

20. Consider a reaction in which malate is oxidized to oxaloacetate by
    malate dehydrogenase in the presence of NAD^{+} which is reduced
    to NADH. These are reactions within the Citric acid cycle or

Krebs cycle. Use $E^{o'}$ values on Table 3.3 to determine the $E^{o'}$ of this reaction and the accompanying free energy change. Is this reaction spontaneous? Explain.

> Answer –0.154 V; 29.7 kJ. With $E^{o'}$ negative and ΔG positive, the reaction is nonspontaneous.

PHYSICAL CONSTANTS

Quantity	Symbol	Value
Speed of light	c	2.99792458×10^8 m/s
Elementary charge	e	1.602177×10^{-19} C
Faraday constant	$F = N_A e$	9.64853×10^4 C/mol
Boltzmann constant	k	1.38066×10^{-23} J/K
Gas constant	R	8.31451 J/K-mol
Planck constant	h	6.62608×10^{-34} J-s
Avogadro constant	N_A	6.02214×10^{23}/mol
Mass		
Electron	m_e	9.10939×10^{-31} kg
Proton	m_p	1.67262×10^{-27} kg
Neutron	m_n	1.67493×10^{-27} kg
Vacuum permittivity	$\varepsilon = 1/c^2\mu_o$	8.8519×10^{-12} C^2/J-m
	$4\pi\varepsilon_o$	1.11265×10^{-12} C^2/J-m
Vacuum permeability	μ_o	$4\pi \times 10^{-7}$ J-s^2/C-m
Magneton		
Bohr	$\mu_B = e\hbar/2m_e$	9.27402×10^{-24} J/T
Nuclear	$\mu_N = e\hbar/2m_p$	5.05079×10^{-27} J/T
g value	g_e	2.00232
Bohr Radius	$a_o =$	5.29177×10-11 m
Stefan-Boltzmann constant	σ	5.67051×10^{-8} W/m^2-K^4
Rydberg constant		1.09737×10^5/cm
Gravitational constant	G	6.67259×10^{-11} Nm^2/kg^2

Useful Relations

At 298.15 K

RT	2.4790kJ/mol
RT/F	25.693 mV
RT ln10/F	59.160 mV
KT/hc	207.23/cm
KT/e	25.693 meV

Conversion Factors

1 eV	1.60218×10^{-19} J	$1\ J = 10^{7}$ ergs = 0.239 cal
	86.485 kJ/mol	$1\ V = 1J/C$
	8065.5/cm	
1 cal	4.184 J = 101.325 kPa	
	760 Torr	
1/cm	1.9864×10^{-23} J	
1 Å	10^{-10} m = 1×10^{-8} cm = 10×10^{-10} m	
1D	3.33564×10^{-30} C-m	
1 atm	760 torr = 14.696 psi 1 torr = 1mm Hg	

Prefixes

z	a	f	p	n
zepto	**atto**	**femto**	**pico**	**nano**
10^{-21}	10^{-18}	10^{-15}	10^{-12}	10^{-9}

μ	m	c	d	Da
Micro	**milli**	**centi**	**deci**	**deka**
10^{-6}	10^{-3}	10^{-2}	10^{-1}	10^{1}

k	M	G	T
Kilo	**mega**	**giga**	**tera**
10^{3}	10^{6}	10^{9}	10^{12}

The Greek Alphabets

A, α	alpha	N,ν	nu
B,β	beta	Ξ,ξ	xi
Γ,γ	gamma	O,o	omicron
Δ,δ	delta	Π,π	pi
E,ε	epsilon	P,ρ	rho
Z,ζ	zeta	Σ,σ	sigma
H,η	eta	T,τ	tau
Θ,θ	theta	Y,υ	upsilon
I,ι	iota	Φ,φ	phi
K,κ	kappa	X,χ	chi
Λ,λ	lambda	Ψ,ψ	psi
M,μ	mu	Ω,ω	omega

95

About the Author

Maurice O. Iwunze obtained a Ph.D. degree in chemistry from Baylor University. While there he worked on the mechanisms of electrochemical oxidation of organic compounds, solubilized in micelles and microemulsion systems. After working briefly at the Federal University of Technology, Yola, Nigeria, he went to the University of Connecticut, Storrs as a research fellow working on advanced electrochemical techniques in bicontinuous microemulsion. He has been on the faculty of the department of chemistry at Morgan State University since 1990 where he continues to do research in the area of electrochemical application in the detoxification of organic pollutants and electron transfer reactions in organized media including sol-gel matrices.